THE LAPLACE TRANSFORM

The Laplace Transform

Richard E. Bellman
Professor of Mathematics,
Electrical Engineering and Medicine
University of Southern California
Los Angeles, California 90007

Robert S. Roth
The Charles Draper Laboratory, Inc.
Cambridge, Massachusetts 02139

World Scientific

Published by

World Scientific Publishing Co Pte Ltd.
P O Box 128, Farrer Road, Singapore 9128.

QA
432
.B45
1984

THE LAPLACE TRANSFORM
Copyright © 1984 by World Scientific Publishing Co Pte Ltd.

ISBN 9971-966-73-5

Printed in Singapore by Richard-Clay (S.E. Asia) Pte. Ltd.

To Nina

PREFACE

The demands of modern science and technology continually force the mathematician to explore new ways to solve problems which are rooted in the ideas of classical mathematics. The computer, first with its gigantic memory and power and now represented by the microprocessor, has subtly changed the way applied mathematicians approach their task of problem solving. While these advances have greatly improved the way mathematics is done, it has led to a trend away from the continual re-evaluation of classical mathematics to a more pragmatic approach for problem solving.

While we applaud this modern approach, we, nevertheless, feel that a blend of classical and modern techniques serves best to enrich the appreciation of the problems under consideration. Furthermore, sometimes seeking the most interesting solution to a problem leads to surprisingly new ideas.

The classical theory of the Laplace Transform can open many new avenues when viewed from a modern semi-classical point of view. In this book we intend to re-examine the Laplace Transform and in so doing we will study many of the applications which makes the Laplace Transform a truly modern tool whose roots are steeped in classical traditions.

In Chapter 1 we shall examine the Laplace Transform in

terms of its classical definition. Of particular interest are the general conditions under which the Laplace Transform of a function u(t) exists. While questions of this type are important in classical analysis, when a numerical procedure is used to obtain a solution, such *a priori* knowledge of its behavior is crucial to obtaining an accurate solution in an economical way.

The Stieltjes integral theory provides a convenient framework within which the Laplace Transform can be developed and suggests ways in which the transform can be used in unconventional ways. Since the Laplace Transform is defined in terms of a complex parameter, s, we have at our disposal the rich theory of complex variables which we use to the fullest to obtain the inverse transform with all its interesting properties.

Chapter 2 is concerned with using the Laplace Transform to solve problems governed by linear ordinary differential equations. An interesting observation concerning these problems is that quite a bit of information about the solution can be obtained from the Laplace Transform even before an inversion is attempted. Similar results are obtained for systems of linear ordinary differential equations. Finally, as we shall show at the end of the chapter, the Laplace Transform can be used to study the behavior of higher derivatives without explicitly determining the basic solution.

Chapter 3 applies the Laplace Transform to differential-difference equations of the form

$$(1) \qquad u'(t) = u(t-1)$$

which form a very interesting class of differential equations. Such equations serve as models for biological systems with inherent delay mechanisms and practical problems of feedback control where delays are the result of design considerations.

Mathematically such equations provide us with very

challenging problems. We begin the chapter by asking a fundamental
question: what is the correct formulation of the initial value
problem? An elementary investigation yields rather startling
results. Initially equations of the form (1) are begun, not by a
definition of the state of the system at a particular time, but
rather the system is begun by defining its behavior over an initial
time interval. With this our interest is stimulated by the
additional flexibility in moulding such equations to individual
problems.

In investigating existence and uniqueness of the solutions
to the differential-difference equations, whole classes of equations become apparent. Laplace Transform techniques can be
successfully applied to these equations with results leading to
the solution of transcendental characteristic equations whose
structure determines the general solution of the ultimate solution.

A close examination of the differential-difference equation
shows that it can be linked to an ordinary differential equation
thru a limiting process, leading us to the observation that in
certain instants a more accurate description of a phenomenon yields
an easier problem while the obvious approximation presents a more
difficult one.

Chapter 4 is a study of the three fundamental partial
differential equation of mathematical physics. These equations,
in which the dimensionality of the system may increase several fold,
presents many problems to the analyst. Numerical solutions to
partial differential equations are particularly troublesome. As
we point out in this chapter, the Laplace Transform technique
serves to reduce the dimension of the problem at least by one by
replacing a variable of the system by a complex parameter.

The situation presented to the analyst in the solution of
these three types of partial differential equations is that of
choosing among many different techniques those which best serve
their purpose. In this chapter we examine these concepts.

The renewal equation, which we consider in detail in Chapter 5, is a linear integral equation of the form

$$(2) \qquad a(t) + \int_0^t g(t-s) f(s)ds = f(t)$$

defining the behavior of $f(t)$ in terms of a displacement kernel $g(t)$ and a forcing function $a(t)$. By considering the formal Laplace Transform technique, the solution can be expressed in an interesting form. We shall examine some of the conditions under which the Laplace Transform technique can be successfully applied to the renewal equation.

As we show, the advantage of the Laplace Transform is that it permits us to predict some of the important behavior of the solution before the actual solution is found. Such information is valuable when numerical solutions are required which brings us directly to Chapter 6.

In Chapter 6 we confront the problem of the numerical inversion of the Laplace Transform. Classically the power of the transform techniques revolved around the simple analytical expressions which were available for the expression of the inverse. This, however, restricted the class of problems which could be solved by this method. Our vision has been greatly expanded thru the use of large scale computers. As a result we are far less timid in our choice of difficult problems.

In this chapter we shall explore one method of obtaining the inverse transform by direct computation. The basic route we will follow is to use the method of numerical quadrature to reduce the integral equation,

$$(3) \qquad F(s) = \int_0^\infty e^{-st} f(t)dt$$

to a system of algebraic equations. A detailed examination of these techniques will reveal several numerical problems which, as we will show, will require sophisticated ideas.

The mathematician is a very curious fellow, always seeking new and interesting ways of solving old and new problems. In this book, we have re-examined some of the elementary results from the classical theory of the Laplace Transform to see how they apply to not only well-known differential equations but also to differential-difference equations and the renewal equation. Our aim has not been to present an exhaustive investigation of the Laplace Transform, but rather to follow paths which were interesting. Much material has not been considered and many paths have been left unexplored. The reader, being also curious, is encouraged to branch out on his own. New ideas, based on sound mathematical foundations, are always needed and their discoveries can only lead to a deeper understanding on an unknown multitude of selected areas of mathematics.

Santa Monica, California   Richard E. Bellman

January 1, 1982   Robert S. Roth

CONTENTS

Chapter 1

THE LAPLACE TRANSFORM

1.1   Introduction

The Laplace Transform is a very neat mathematical method
for solving problems which arise in several areas of mathematical
analysis. Of particular importance is its ability to solve
differential equations, partial differential equations and
differential-difference equations which continually arise in
engineering problems.

In this chapter we will review the formal mathematical
definition of the Laplace Transform and derive some of its fundamental properties. Our interest in the Laplace Transform is quite
practical. We wish to solve problems and our experience has shown
us that the Laplace Transform is a valuable tool in doing so. We
wish to point out also that the use of the Laplace Transform
increases in power when it is used in somewhat unconventional ways.

With the availability of modern large scale computers, its
applications have become increasingly important as a tool in the
numerical solution of mathematical problems. With this clear understanding the analyst and engineer must, in dealing with today's
broad spectrum of technical problems, consider both analytical and
numerical properties of the Laplace Transform.

Because of our interest in examining unconventional

applications of the Laplace Transform, we shall consider further the concept of the Stieltjes integral which allows us to extend the ideas of the transform to problems involving discontinuities. To begin, let $u(t)$ be a continuous real function defined for all $t \geq 0$. The function $L(u)$, given by the expression,

$$(1.1.1) \qquad L(u) = \int_0^\infty e^{-st} u(t)dt = F(s)$$

is defined as the Laplace Transform of u. Traditionally the function is written either as $L(u)$, to show its dependence on the function $u(t)$, or as $F(s)$, indicating that (1.1.1) is a function of the complex variable $s = (x + iy)$.

Before we examine some of the properties of the Laplace Transform of $u(t)$, we must pay careful attention to convincing ourselves that the integral (1.1.1) does, in fact, exist. To do this, we shall construct a rather general form of the integral, the Stieltjes integral, and show that it includes, as a subcase, the conventional definition of the Laplace Transform.

1.2   Functions of a Bounded Variation

The function $F(s)$ exists whenever the integral,

$$(1.2.1) \qquad \int_0^\infty e^{-st} u(t)dt = F(s)$$

exist over a range of the complex parameter s. Clearly this puts some restrictions on the behavior of $u(t)$, particularly as t gets very large.

To broaden our vision as much as possible, let us consider some of the generalizations which can be made under which (1.2.1) exists. To this end we introduce the concept of a function of a bounded variation. Such a function $f(t)$ is said to be of a bounded variation in the interval (a, b) if it can be expressed in the form $g(t) - h(t)$ where both functions $g(t)$ and $h(t)$ are nondecreasing bounded functions.

1.3   The Stieltjes Integral

The integral defined in (1.1.1) is the usual Riemann integral used for the conventional definition of the Laplace Transform. Yet to make sure we will be working in a mathematical frame which is broad enough to include a large variety of applications, we shall introduce the Stieltjes integral as a generalization of the more widely used Riemann integral. Our aim in constructing this integral is to demonstrate the general conditions under which the Laplace Transform exists.

To construct the Stieltjes integral, let $\alpha(x)$ and $f(x)$ be real bounded functions of a real variable x for $a \leq x \leq b$. Define a subdivision A of the interval (a, b) by the points,

$$a = x_0 < x_1 < x_2 \ldots < x_n = b \quad .$$

Let δ be the largest $|x_{i+1} - x_i|$, $\quad i = 0,1,\ldots,n-1$. If

$$\lim_{\delta \to 0} \sum_{i=0}^{n-1} f(\xi_i) \, (\alpha(x_{i+1}) - \alpha(x_i))$$

$$x_i < \xi_i < x_{i+1}$$

exists independently of the manner of the subdivision and the choice of ξ_i, then the limit is called the Stieltjes integral of $f(x)$ with respect to $\alpha(x)$ and is denoted,

$$(1.3.1) \qquad \int_a^b f(t) \, d\alpha(t)$$

Using the definitions we have just introduced, it can be shown to be true that if $f(x)$ is continuous and $\alpha(x)$ is of bounded variation in the interval (a, b), then the Stieltjes integral of $f(x)$ with respect to $\alpha(x)$ space from a to b exists.

A further characterization of the existence of the Stieltjes integral is given by the following result. If $f(x)$ and $\alpha(x)$ are real bounded functions in $a < x < b$ and in addition $\alpha(x)$ is nondecreasing, then the necessary and sufficient condition that,

$$\int_a^b f(x) \, d\alpha(x)$$

exists, is that

$$\lim_{\delta \to 0} (S_A - s_A) = 0 \quad ,$$

independent of the manner of subdivision, where

$$S_A = \sum_{k=0}^{n-1} M_k (\alpha_{k+1}(x) - \alpha_k(x))$$

$$s_A = \sum_{k=0}^{n-1} m_k (\alpha_{k+1}(x) - \alpha_k(x))$$

$$M_k = \text{l.u.b. } f(x)$$

$$m_k = \text{g.l.b. } f(x)$$

$$x_k \leq x \leq x_{k+1}$$

By using these results, we have a means of establishing the existence of the Stieltjes integral when the need arises.

1.4 Improper Stieltjes Integral

If we recall that the Laplace Transform $L(u)$ defined in (1.1.1) has a range of integration from 0 to infinity, we inquire what the counterpart in the framework of the Stieltjes integral would be. The improper Stieltjes integral can be defined as a limiting process in the following way. If $f(x)$ is continuous in $a < x < R$, for every R, then

$$(1.4.1) \qquad \int_a^\infty f(x)\ d\alpha(x) = \lim_{R \to \infty} \int_a^R f(x)\ d\alpha(x) \quad ,$$

when the limit exists. If the integral does, in fact, exist, the improper integral (1.4.1) is said to converge. Furthermore, we say (1.4.1) converges absolutely if and only if,

$$(1.4.2) \qquad \int_a^\infty |f(x)|\ dV(x) < \infty$$

where $V(x)$ is associated with the bounded variation property of $\alpha(x)$ and is defined as,

$$(1.4.3) \qquad V(x) = g(x) - h(x) \quad .$$

where both $g(x)$ and $h(x)$ are nondecreasing bounded functions.

1.5 The Laplace Transform

To demonstrate the conditions under which the Laplace Transform exists, we shall consider the results we have found for the Stieltjes integral. Recall from the last section, that if $f(x)$ is continuous and $\alpha(x)$ is of bounded variation, and,

$$(1.5.1) \qquad \int_0^\infty |f(x)|\ dV(x) < \infty \quad ,$$

then the integral

$$(1.5.2) \qquad \int_0^R f(x)\ d\alpha(x)$$

exists and in the limit as $R \to \infty$ converges absolutely.

Now, if $\alpha(x)$ is a continuous function of a nondecreasing parameter t, then we can write $\alpha(x) = t$ so that $d\alpha(x) = dt$, and

$$(1.5.3) \qquad f(x) = f(t) = e^{-st}\ u(t) \quad ,$$

where s is a complex parameter.
Then, the Stieltjes integral (1.4.1) becomes the Laplace Transform,

$$(1.5.4) \qquad \int_0^\infty e^{-st} u(t)dt \quad .$$

Now if we impose the bounded condition on the function $u(t)$,

$$(1.5.5) \qquad |u(t)| < ae^{bt} \quad ,$$

for some constants a and b as $t \to \infty$, and that

$$(1.5.6) \qquad \int_0^T |u(t)|dt < \infty$$

for any finite T, then the conditions are satisfied to insure that the integral will converge absolutely and uniformly for $Re(s) > b$, since

$$\int_0^\infty |e^{-st} u(t)|dt \leq a\int_0^\infty |e^{(b-s)t}|dt < \infty$$

for $Re(s) > b$.

Therefore, depending on the bounded behavior of $u(t)$ for large t, a convergent Laplace Transform exists only for those complex parameters s whose real parts are greater than b.

1.6   Existence and Convergence

The explicit results of the last section may be restated in the following way. If $u(t)$ is continuous and satisfies a bound of the form,

$$(1.6.1) \qquad |u(t)| \leq ae^{bt}$$

for some constants a and b as $t \to \infty$ and if

$$(1.6.2) \qquad \int_0^T |u(t)|dt < \infty$$

for every finite T, then the combination of these two reasonable assumptions permits us to conclude that the integral,

(1.6.3) $\quad \int_0^\infty e^{-st} u(t)dt$

exists and converges absolutely and uniformly for $\text{Re}(s) > b$.

1.7 Properties of the Laplace Transform

Some of the elementary properties of the Laplace Transform are of particular interest to us as we proceed through the book. We shall begin with the most important of the elementary attributes of the Laplace Transform, namely the relative invariance under translations in both the t- and s-spaces.

We first observe that from its definition, the Laplace Transform can be written,

(1.7.1) $\quad \int_0^\infty e^{-st} e^{-bt} u(t)dt = \int_0^\infty e^{-(s+b)t} u(t)dt \quad .$

Or in other words,

(1.7.2) $\quad L(e^{-bt}u(t)) = F(s+b) \quad .$

Also we have, using the same technique of examination,

(1.7.3) $\quad \int_1^\infty e^{-st} u(t-1)dt$

$$= \int_0^\infty e^{-s(t+1)} u(t)dt = e^{-s} L(u) \quad .$$

Another property of the Laplace Transform can be illustrated by considering,

(1.7.4) $\quad L(du/dt) = \int_0^\infty e^{-st} (du/dt)dt$

$$= (e^{-st} u(t)) \Big|_0^\infty + s\int_0^\infty e^{-st} u(t)dt \quad ,$$

by integrating by parts.
From the convergence requirements of the Laplace Transform, we

assume that,

(1.7.5) $\quad \lim_{t \to \infty} e^{-st} u(t) = 0$.

Using these results we obtain the property that,

(1.7.6) $\quad L(du/dt) = sL(u) - u(0)$.

Similarly,

(1.7.7) $\quad L(d^2u/dt^2) = sL(du/dt) - u'(0)$

$$= s^2 L(u) - su(0) - u'(0) \quad ,$$

and inductively,

(1.7.8) $\quad L(d^n u/dt^n) = s^n L(u) - s^{n-1}u(0) - \ldots - u^{(n-1)}(0) \quad ,$

where,

$$u^{(n)}(x) = d^n u/dx^n \quad .$$

This is a remarkable property of the Laplace Transform for it transforms derivatives into simple algebraic expressions together with the inclusion of initial conditions in a very natural way.

The Laplace Transform has several other far reaching properties. For example, if $f(t)$ is Laplace Transformable, that is, its Laplace Transform exists, then we can show that

(1.7.9) $\quad L\left(\int_0^t f(t)dt\right) = F(s)/s$.

Hence, we follow the formal definition of the Laplace Transform,

(1.7.10) $\quad L\left(\int_0^t f(t)dt\right) = \int_0^\infty e^{-st} \int_0^t f(p)dp\,dt$.

Integrating the right side of (1.7.10) by parts, we have,

$$(1.7.11) \quad L\left(\int_0^t f(t)dt\right) = -\frac{1}{s} e^{-st} \int_0^t f(p)dp \Big|_0^\infty + \frac{1}{s}\int_0^\infty e^{-st} f(t)dt$$

$$= F(s)/s.$$

Using the same technique, we can obtain two interesting similar results.

$$(1.7.12) \quad L(t\, f(t)) = \int_0^\infty e^{-st} t\, f(t)dt$$

$$= -\int_0^\infty \frac{d}{ds}\,(e^{-st}\, f(t))dt$$

$$= -\frac{dF(s)}{ds}\,.$$

We also take note of the following fact,

$$(1.7.13) \quad L(f(t)\,/\,t)) = \int_0^\infty e^{-st}\, f(t)/t\ dt$$

$$= \int_0^\infty \int_0^s e^{-pt}\, f(t)dpdt$$

$$= \int_0^s \int_0^\infty e^{-pt}\, f(t)dtdp$$

$$= \int_0^s F(p)dp\quad .$$

Finally, we wish to examine the Laplace Transform as it defines a complex function over a complex domain. By definition, if $u(t)$ is continuous for $t > 0$, then we have defined,

$$(1.7.14) \quad L(u) = \int_0^\infty e^{-st}\, u(t)dt\quad ,$$

where now we wish to emphasize that s is a complex number, i.e.,

$$(1.7.15) \quad s = x + iy$$

where the symbol i is the imaginary number $i = \sqrt{-1}$.

Recalling that,

(1.7.16) $e^{-st} = e^{-(x+iy)t} = e^{-xt} (\cos(yt) - i \sin(yt))$,

we see that the transform (1.7.14) can be written,

(1.7.17) $L(u) = \int_0^\infty e^{-xt} (\cos(yt) - i \sin(yt)) \, u(t)dt$.

By inspection we can write,

(1.7.18) $L(u) = F(x,y) + i \, G(x,y)$,

demonstrating clearly that $L(u)$ is a complex function of a complex variable. Quite often problems are encountered in which the unit step function plays a prominent part. This function is usually denoted as $U(t)$ and its definition is

(1.7.19) $U(t) = \begin{cases} 0 & t < 0 \\ \\ 1 & t \geq 0 \end{cases}$.

We now have the following results, sometimes called the shifting theorem. The theorem states that if

$$L(f) = F(s)   ,$$

then

$$L(f(t-T) \, U(t-T)) = e^{-sT} F(s)   .$$

We demonstrate the proof of this result by noting that,

(1.7.20) $L(f(t-T) \, U(t-T)) = \int_0^\infty f(t-T) \, U(t-T) \, e^{-st} \, dt$

$$= \int_T^\infty f(t-T) \, e^{-st} \, dt$$

by letting $p = t-T$,

$$= \int_0^\infty f(p) \, e^{-s(p+T)} \, dp$$

$$= e^{-st} \int_0^\infty f(p) \, e^{-sp} \, dp$$

$$= e^{-sT} F(s) \quad .$$

1.8   The Inversion of the Laplace Transform

The Laplace Transform of a function $u(t)$ is of little value to the analyst if there is no assurance that the inverse exists and is unique. Therefore we shall be concerned in this section with demonstrating that such an inverse exists uniquely and in so doing we will determine the conditions under which the inversion can be successfully performed.

Since we are not interested in the most general case, but only in the class of functions which arise naturally in the course of our investigations, we shall restrict ourselves to proving the following: If $F(s)$ satisfies the following conditions,

1. $F(s)$ is analytic for $\text{Re}(s) > a$

2. $F(s) = \dfrac{c_i}{s} + 0(\dfrac{1}{|s|^2})$ as $|s| \to \infty$ along $s = b + it$

$$b > a \quad ,$$

then

$$(1.8.1) \qquad f(t) = \frac{1}{2\pi i} \int_{(c)} F(s) \, e^{st} \, ds \quad , \qquad t > 0 \quad ,$$

where (c) is a contour in the region of analyticity, exists and

$$(1.8.2) \qquad F(s) = L(f) \quad .$$

To demonstrate the result is true, let us write the Laplace Transform of a function $f(t)$, $t \geq 0$, as follows,

$$(1.8.3) \qquad L(f) = \lim_{T \to \infty} \int_0^T f(t) \, e^{-pt} \, dt \quad ,$$

where p is a complex parameter.

Using (1.8.1) we have

$$(1.8.4) \qquad L(f) = \lim_{T \to \infty} \int_0^T \left[\frac{1}{2\pi i} \int_{(c)} F(s) e^{st} ds \right] e^{-pt} dt \quad .$$

Because of the absolute convergence of the double integral, the order of integration can be reversed, and we have

$$(1.8.5) \qquad L(f) = \lim_{T \to \infty} \frac{1}{2\pi i} \int_{(c)} F(s) \int_0^T e^{(s-p)t} dt ds \quad .$$

If $Re(p) > Re(s)$, the inner integral exists in the limit as $T \to \infty$ and may be explicitly evaluated.
Therefore,

$$(1.8.6) \qquad L(f) = \lim_{T \to \infty} \frac{1}{2\pi i} \int_{(c)} F(s) \left[\frac{e^{(s-p)T} - 1}{(s-p)} \right] ds \quad ,$$

or in the limit as $T \to \infty$

$$(1.8.7) \qquad L(f) = \frac{-1}{2\pi i} \int_{(c)} \frac{F(s)\ ds}{(s-p)} = F(p)$$

because $F(s)$ is analytic for $Re(s) > a$.

Alternatively by direct manipulation, we can show the existence of the inverse Laplace Transform by the following analysis. If $f(t)$ is defined for $t > 0$ and the correspond transform,

$$(1.8.8) \qquad F(s) = \int_0^\infty f(t)\ e^{-st}\ dt$$

is absolutely convergent for $Re(s) > a$, as in Fig. 1.1.
Then for $x > a$, we can write,

$$F(x + iy) = \int_0^\infty f(t)\ e^{-(x+iy)t}\ dt \quad ,$$

where the integral is absolutely convergent.

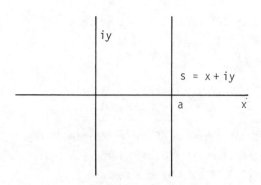

Fig. 1.1 Region of absolute convergence for
the Laplace Transform

Now, if we multiply both sides of (1.8.8) by $e^{u(x+iy)}$,
u being a real parameter, and integrate between -T and T,
along the imaginary y-axis, we see,

$$(1.8.9) \qquad \int_{-T}^{T} e^{u(x+iy)} F(x+iy)dy$$

$$= \int_{-T}^{T} e^{u(x+iy)} \int_{0}^{\infty} e^{-(x+iy)t} f(t)dtdy$$

But because the double integral converges absolutely, the order of
integration can be interchanged,

$$(1.8.10) \qquad \int_{-T}^{T} e^{u(x+iy)} F(x+iy)dy$$

$$= e^{ux} \int_{0}^{\infty} f(t)e^{-xt} \int_{-T}^{T} e^{(iuy-iyt)}dydt \qquad .$$

We can now note that

$$(1.8.11) \qquad e^{i(u-t)y} = \cos(u-t)y + i\ \sin(u-t)y \qquad ,$$

and therefore

$$(1.8.12) \qquad \int_{-T}^{T} e^{i(u-t)y}\ dy = \frac{2\sin T(u-t)}{(u-t)} \qquad .$$

Substituting (1.8.12) into (1.8.10), we see

$$(1.8.13) \quad \int_{-T}^{T} e^{u(x+iy)} F(x+iy) dy$$

$$= 2e^{ux} \int_{0}^{\infty} f(t) e^{-xt} \frac{\sin T(u-t)}{(u-t)} dt \quad .$$

To investigate the behavior of the right-hand integral for $u > 0$, we can break the interval $(0, \infty)$ into $(0, u-d)$, $(u-d, u+d)$, $(u+d, \infty)$, so (1.8.13) can be written as,

$$(1.8.14) \quad \int_{-T}^{T} e^{u(x+iy)} F(x+iy) dy$$

$$= 2e^{ux} \left[\int_{0}^{(u-d)} (.) dt + \int_{(u-d)}^{(u+d)} (.) dt + \int_{(u+d)}^{\infty} (.) dt \right] \quad .$$

Each integral of (1.8.14) is of the form,

$$\int_{a}^{b} g(t) \left\{ \begin{matrix} \sin(tT) \\ \cos(tT) \end{matrix} \right\} dt$$

where $g(t)$ is continuous over the intervals defined on the first and third terms. If we assume that $g(t)$ has a derivative, then integrating by parts gives, for example,

$$\int_{a}^{b} g(t) \sin(tT) dt$$

$$= \frac{g(t) \cos(tT)}{T} \Big|_{a}^{b} - \frac{1}{T} \int_{a}^{b} g'(t) \cos(tT) dt$$

and we have the result,

$$\lim_{T \to \infty} \int_{a}^{b} g(t) \left\{ \begin{matrix} \sin(tT) \\ \cos(tT) \end{matrix} \right\} dt = 0 \quad ,$$

therefore in the limit as $T \to \infty$, the first and third integrals of (1.8.14) vanish.

Therefore, at this point we can say, in the limit,

(1.8.15)
$$\int_{-T}^{T} e^{u(x+iy)} F(x+iy)dy$$

$$= 2e^{ux} \int_{(u-d)}^{(u+d)} f(t)e^{-xt} \frac{\sin T(u-t)}{(u-t)} dt \quad .$$

Since d is small, we assume f(t) is sufficiently smooth in the neighborhood of u that,

$$f(t)e^{-xt} = f(u)e^{-xu} + h(u, t)(t-t) \quad ,$$

where $|h(u, t)| < k$, $(u-d) < t < (u+d)$.
Therefore, we can immediately write,

(1.8.16)
$$\int_{-T}^{T} e^{u(x+iy)} F(x+iy)dy$$

$$= 2f(u)\int_{(u-d)}^{(u+d)} \frac{\sin T(u-t)}{(u-t)} dt$$

$$+ 2e^{xu} \int_{(u-d)}^{(u+d)} h(u, t) \sin T(u-t)dt \quad .$$

Since $|\sin T(u-d)| < 1$ and $|h(u, t)| < k$, the second integral is $O(d)$. Now set $v = T(u-t)$, $dv = -Tdt$, we have then,

(1.8.17)
$$\int_{-T}^{T} e^{u(x+iy)} F(x+iy)dy$$

$$= - 2f(u) \int_{-Td}^{Td} \frac{\sin v}{v} dv + O(d)$$

We can use the fact that

$$\int_{-\infty}^{\infty} \frac{\sin x}{x} dx = \pi \quad .$$

If we let $T \to \infty$ and $d \to 0$ such that $Td \to \infty$, then we have

$$(1.8.18) \qquad \lim_{T \to \infty} \int_{-T}^{T} e^{u(x+iy)} F(x+iy)dy = -2\pi f(u) + O(d) \qquad ,$$

and therefore,

$$(1.8.19) \qquad f(u) = \lim_{T \to \infty} -\frac{1}{2\pi} \int_{-T}^{T} e^{u(x+iy)} F(x+iy)dy \qquad .$$

Since $F(s)$ is analytic for $Re(s) > a$, the integral (1.8.19) exists in the right half plane, $Re(s) > a$. Since $s = x+iy$, we can choose a path c in the complex plane parallel to the imaginary axis so that $ds = -idy$. The limits of integration of (1.8.19) become $a \pm iT$.

The final expression can then be written,

$$(1.8.20) \qquad f(u) = \lim_{T \to \infty} \frac{1}{2\pi i} \int_{a-iT}^{a+iT} e^{us} F(s)ds \qquad ,$$

which is the expression for the inverse Laplace Transform.

1.9   The Convolution Theorem

A fundamental property of the Laplace Transform is associated with the expression

$$(1.9.1) \qquad h(t) = \int_{0}^{t} f(r) \, g(t-r)dr \qquad .$$

This mathematical operation creates a function $h(t)$ as a composite of two functions $f(t)$ and $g(t)$ and plays important roles in analysis in mathematical physics and probability theory. The notation,

$$(1.9.2) \qquad h = f * g$$

is frequently used to symbolize (1.9.1) and the integral itself is the convolution of f and g. If we consider the Laplace Transform of $h(t)$ as a limiting process, we can write,

$$(1.9.3) \qquad \int_0^T h(t)\, e^{-st}\, dt = \int_0^T e^{-st} \int_0^t f(r)\, g(t-r)\, dr\, dt \quad ,$$

as $T \to \infty$.

Now consider the repeated integral as a double integral over a region S in Fig. 1.2,

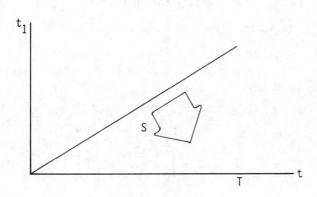

Fig. 1.2

Inverting the order of integration, we have,

$$(1.9.4) \qquad \iint_S e^{-st} f(t_1)\, g(t-t_1) dt_1 dt$$

$$= \int_0^T f(t_1) \left[\int_{t_1}^T e^{-st} g(t-t_1) dt \right] dt_1$$

$$= \int_0^T e^{-st_1} f(t_1) \left[\int_0^{T-t_1} e^{-su} g(u) du \right] dt_1 \quad .$$

As $T \to \infty$, we obtain formally $L(f*g) = L(f)L(g)$. To put this heuristic result on a rigorous foundation, we shall prove the following theorem.

If

$$(a) \qquad \int_0^\infty e^{-at_1} |f(t_1)| dt_1 < \infty$$

(b) $\displaystyle\int_0^\infty e^{-(a+it)t_1} g(t_1)dt_1$ converges for $t \geq 0$,

then

$$\int_0^\infty h(t)e^{-st} \, dt = \left[\int_0^\infty e^{-st} f(t)dt\right]\left[\int_0^\infty e^{-st} g(t)dt\right]$$

for $s = s + ib$, and generally for $\text{Re}(s) > a$.

To begin the proof, we have, referring to (1.9.4),

$$(1.9.5) \quad \int_0^T h(t)e^{-st} \, dt = \int_0^T e^{-st_1} f(t_1)\left[\int_0^{T-t_1} e^{-su} g(u)dt\right]dt_1$$

$$= \int_0^T e^{-st_1} f(t_1)\left[\int_0^\infty e^{-su} g(u)du\right]dt_1$$

$$- \int_0^T e^{-st_1} f(t_1)\left[\int_{T-t_1}^\infty e^{-su} g(u)du\right]dt_1 ,$$

The second integral in (1.9.5) can be shown to be bounded. We break up the range of t_1 integration into two parts, $(0, R)$ and (R, T). Since, by assumption the integrals $\displaystyle\int_0^\infty e^{-su} g(u)du$ and $\displaystyle\int_0^\infty e^{-at}|f(t)|dt$ converge, then for any $\varepsilon > 0$, there is an R, depending on ε for which,

(a) $\displaystyle\left|\int_{R(\varepsilon)}^\infty e^{-su} g(u)du\right| \leq \varepsilon$

(b) $\displaystyle\left|\int_{R(\varepsilon)}^\infty e^{-st} f(t)dt\right| \leq \varepsilon c_2$

(c) $\displaystyle\left|\int_0^\infty e^{-st} g(u)du\right| < c_1$.

Therefore for $R(\varepsilon)$ selected to satisfy (a), we can say

$$(1.9.6) \quad \left| \int_0^{R(\varepsilon)} e^{-st_1} f(t_1) \left[\int_{R(\varepsilon)-c_1}^{\infty} e^{-su} g(u)du \right] dt_1 \right|$$

$$\leq \varepsilon \int_0^{R(\varepsilon)} e^{-at_1} |f(t_1)| dt_1$$

$$\leq \varepsilon \int_0^{\infty} e^{-at_1} f(t_1) dt_1$$

$$\leq \varepsilon c_2 \quad .$$

The remaining integral has the bound,

$$(1.9.7) \quad \left| \int_{R(\varepsilon)}^T f(t_1) \left[\int_{R(\varepsilon)-t_1}^{\infty} e^{-su} g(u)du \right] dt_1 \right|$$

$$\leq c_1 \int_{R(\varepsilon)}^T e^{-at_1} |f(t_1)| dt_1$$

$$< c_1 \varepsilon \quad .$$

By letting $T \to \infty$, the second integral in (1.9.5) goes to zero and we are left with the result,

$$(1.9.8) \quad \int_0^{\infty} h(t) e^{-st} dt = \int_0^{\infty} e^{-st} f(t)dt \int_0^{\infty} e^{-su} g(u)du$$

i.e.,

$$(1.9.9) \quad L(f \star g) = L(f)L(g) \quad .$$

Result (1.9.9) informs us that rather than constructing the Laplace Transform for a complicated convolution involving both $f(t)$ and $g(t)$, we need only construct the transform of each separately and multiply. This device will be used later with the renewal equation.

1.10   Instability of the Inverse of the Laplace Transform

Let $v(t)$ be a function for which

$$(1.10.1) \qquad \int_0^\infty |v(t)| \, e^{-kt} \, dt < \epsilon \quad .$$

Then, for $Re(s) \geq k$,

$$(1.10.2) \qquad |L(u+v) - L(u)| = \left| \int_0^\infty v(t) \, e^{-st} \, dt \right|$$

$$< \int_0^\infty |v(t)| e^{-Re(s)t} < \epsilon \quad .$$

In other words, a "small" change in u produces an equally small change in $L(u)$. In mathematical terms, $L(u)$ is stable under perturbations of this type.

The impossibility of usable universal algorithms for inverting the Laplace Transform is a consequence of the fact that the inverse of the Laplace Transform is not stable under reasonable perturbations. Two simple examples illustrate this. Consider first the well-known formula,

$$(1.10.3) \qquad L(\sin at) = a/(s^2 + a^2) \quad .$$

As a increases, the function sin at oscillates more and more rapidly, but remains of constant amplitude. The Laplace Transform, however, is uniformly bounded by $1/a$ for $s \geq 0$ and thus approaches 0 uniformly as $a \to \infty$.

As a second example we consider the formula,

$$(1.10.4) \qquad L(u) = L\left(\frac{a}{2\sqrt{\pi}} \frac{e^{-a^2/4t}}{t^{3/2}} \right) = e^{-a\sqrt{s}} \quad .$$

As $a \to 0$, the function $e^{-a\sqrt{s}}$ remains uniformly bounded by 1 for $s \geq 0$. Observe how $u(t)$ behaves as a function of t. At $t = a^2/4$, we see that it has the value c_1/a^2 where c_1 is a positive constant. Nonetheless, at $t = 0$, for all $a > 0$, $u(t)$ assumes the value 0. Hence $u(t)$ has a "spike" form (see Fig. 1.3), one which is sharper and sharper as $a \to 0$. We see then that $u(t)$ is

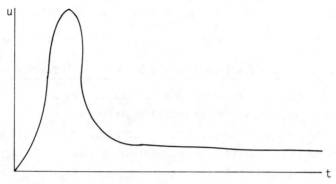

Fig. 1.3

is highly localized in the vicinity of $t = 0$ for a small, and
thus that $u(t)$ is an excellent approximation to the delta
function $\delta(t)$.

These examples make evident some of the difficulties we
face in finding $u(t)$ given $F(s)$. Let $F(s)$ be calculated to
an accuracy of 1 in 10^{10}, say to ten significant figures, then
if $u_0(t)$ is the function giving rise to $F(s)$ via $F(s) = L(u_0)$,
we see that

$$(1.10.5) \qquad u_0 + \sin 10^{20}(t-t_1) + 10^{-20} \; \frac{a_1 e^{-a_1^2 / 4(t-s_1)}}{a \sqrt{\pi} \, (t-s_1)^{3/2}}$$

will have, to ten significant figures, the same Laplace Transform
for any a_1, $s_1 > 0$.

We cannot therefore "filter out" extremely rapid oscillations or spike behavior of $u(t)$ on the basis of numerical values
of $F(s)$ alone. What we can do to escape from this simultaneous
nightmare and quagmire of pathological behavior is to agree to
restrict our attention from the beginning to functions $u(t)$
which are essentially smooth. In other words, we can use knowledge
of the structural behavior of $u(t)$ to obtain numerical values.
In many cases, as we shall show, the inverse transform must be
accomplished by numerical means and we shall be continually keeping

the analytical and numerical behavior of each problem clearly in mind.

1.11   The Laplace Transform and Differential Equations

An important application of Laplace Transform occur in the solution of ordinary differential equations which are cast in the form of initial value problems. Properties of the Laplace Transform make the transform very appealing as a means of finding solutions provided the inverse transform can be easily found.

Let us now turn to the solution of linear equations with constant coefficients by means of the Laplace Transform. To begin with, consider the first-order scalar equation,

$$(1.11.1) \qquad du/dt = au + v \ , \qquad u(0) = c \qquad .$$

Taking the Laplace Transform of both sides, we have

$$(1.11.2) \qquad \int_0^\infty e^{-st} (du/dt)dt = a \int_0^\infty e^{-st} u(t)dt + \int_0^\infty e^{-st} v(t)dt \quad .$$

Hence by integrating by parts,

$$(1.11.3) \qquad e^{-st} u \Big|_0^\infty + s \int_0^\infty e^{-st} udt = a \int_0^\infty e^{-st} udt + \int_0^\infty e^{-st} vdt \quad .$$

Writing,

$$(1.11.4) \qquad L(u) = \int_0^\infty e^{-st} udt \ , \qquad L(v) = \int_0^\infty e^{-st} vdt \quad ,$$

we have, by solving (1.11.3) for $L(u)$,

$$(1.11.5) \qquad L(u) = \frac{c}{s-a} + \frac{L(v)}{s-a} \qquad .$$

The inverse of the first term is known to be ce^{at}. To obtain the inverse of the second term, we apply the convolution theorem. The result is

$$(1.11.6) \qquad u(t) = ce^{at} + \int_0^t e^{a(t-s)} v(s) \ ds \quad .$$

Turning to the vector-matrix case,

(1.11.7) $dx/dt = Ax + y$, $x(0) = c$,

we have

(1.11.8) $L(x) = (sI - A)^{-1} c + (sI - A)^{-1} L(y)$.

Since the inverse transform of $(sI - A)^{-1}$ is e^{At}, we obtain the expression,

$$(1.11.9)   x = e^{At} c + \int_0^t e^{A(t-s)} y(s) \, ds .$$

1.12   Transient Solutions

A byproduct of the Laplace Transform technique when it is applied to differential equations is the fact that one may be able to get explicit expressions for the transient solutions which reflect the initial conditions. Referring to the last section, we see that the differential equation itself (1.11.1) carries the initial condition, $u(0) = c$, as an auxilary condition. The form of the solution given by the Laplace Transform (1.11.5) shows that the initial condition, c, is incorporated in the explicit analytic expression.

If the transform can be inverted analytically, then the results shows explicitly how the initial conditions are propagated in time. If the transform must be inverted numerically, then one cannot tell for sure how the initial conditions propagate and the numerical solution has to be constructed very carefully.

1.13   Generating Functions

The generating function is an example of a transform on functions of a discrete variable, or index, and is formally quite similar to the Laplace Transform.

If $\{u_n\}$, $n = 1, 2, \ldots$ is a sequence, the generating function associated with set $\{u_n\}$ is defined as,

(1.13.1) $\quad G(z) = \sum_{n=0}^{\infty} u_n z^n$

where z is a complex number. Several examples come immediately to mind,

(a) $u_n = 1, \quad n \geq 0$ $\qquad\qquad$ $G(z) = \sum_{n=0}^{\infty} z^n$,

(b) $u_n = \left\{ \begin{array}{l} 0 \\ 1 \end{array} \right\} \begin{array}{l} 0 \leq n \leq m \\ n \geq m \end{array}$ $\qquad$ $G(z) = \sum_{n=m}^{\infty} z^n$,

(c) $u_n = \left\{ \begin{array}{l} 0 \\ u_{n-m} \end{array} \right\} \begin{array}{l} 0 \leq n \leq m \\ n \geq m \end{array}$ $\qquad$ $G(z) = \sum_{n=m}^{\infty} u_{m-n} z^n$.

The inversion of the transform defined by (1.13.1) can quickly be obtained by two different methods. Noting that $G(z)$ is analytic in z, we obtain by differentiation,

(1.13.2) $\quad u_n = 1/n! \left. \dfrac{d^n G(z)}{dz^n} \right|_{z=0}$.

Relating to the analyticity of $G(z)$ we multiply both sides of (1.13.1) by z^{-m-1}, for a given m. Then

(1.13.3) $\quad z^{-m-1} G(z) = \sum_{n=0}^{\infty} u_n z^{n-m-1}$.

If we integrate (1.13.3) by a contour integral about a simple closed contour including the origin and with the region of analyticity,

(1.13.4) $\quad \displaystyle\int_C z^{-m-1} G(z) dz = 2\pi i u_m$, or,

$$(1.13.5) \qquad u_m = \frac{1}{2\pi i} \int_C z^{-m-1} G(z)dz \quad .$$

The convolution of two sequences can be defined in a way analogous to the convolution for the Laplace Transform. Let $\{u_n\}$ and $\{v_n\}$ be two sequences and defined a third sequence $\{w_n\} = \sum_{i=0}^{n} u_i v_{n-i}$. We then write $\{w_n\} = \{u_n\} * \{v_n\}$ and say that $\{w_n\}$ is the convolution of $\{u_n\}$ and $\{v_n\}$. Under these definitions it is easy to see that the generating function of the convolution of the two sequences is the product of the generating functions of the two sequences.

Problems

1. The Laplace Transform is defined as,

$$F(s) = \int_0^\infty e^{-st} f(t)dt \quad .$$

If s is a complex number $(s = x + iy)$:

(a) Show that $F(s)$ is a complex function of a complex variable.

(b) If $f(t)$ is bounded on $0 < t < \infty$, determine the region of convergence of $F(s)$ in the complex plane. Show $F(s)$ is analytic in this region.

2. Find the Laplace Transforms for the following functions.
   (a) $f(t) = 1$
   (b) $f(t) = t$
   (c) $f(t) = t^n$
   (d) $f(t) = \sin at$
   (e) $f(t) = \cosh at$

(f) $f(t) = e^{-at}$

(g) $f(t) = t \sin at$

(h) $f(t) = U(t)$ where $U(t) = \begin{cases} 1 & \text{for} \quad t > 0 \\ 0 & \text{for} \quad t \leq 0 \end{cases}$.

3. Show that the Laplace Transform of $f(t) = e^{-at} \sin bt$ is
$$F(s) = b/((s+a)^2 + b^2) \quad .$$

   (a) Where in the complex plane is $F(s)$ analytic?

   (b) If $a > 0$, is $f(t)$ stable, (i.e., is it bounded at infinity?) and where are the poles of $F(s)$?

   (c) If $a < 0$, $f(t)$ is unstable. How are the poles of $F(s)$ shifted?

   (d) Given $F(s)$, find $f(t)$ by the inversion formulae.

4. What is the Laplace Transform of $f'(t)$?
   Under what conditions on $f(t)$ does this transform exist?
   What must be specified to make the transform unique?
   Does the transform have a zero or pole in the complex plane and if so where are they located?

5. Repeat (4) for $f''(t)$.

6. Repeat (4) for $\int_0^t f(s)ds$.

7. If $F(s) = F(s-a)$, then what can be said for the function $f(t)$?
   State in words how a linear transformation in the transform will effect the function $f(t)$.
   What happens if a is complex?

8. Given a function of two variables $f(x, y)$, the Laplace Transform is defined to be,

$$F(u, v) = \int_0^\infty \int_0^\infty e^{-ux - vy} f(x, y)dxdy \quad .$$

   Under what conditions does $F(u, v)$ exist?

9. Show $L(f(t) + g(t)) = L(f(t)) + L(g(t))$.

10. Using (9) and (10), explain why $L(f)$ is a linear operator. Why is this an important observation?

Chapter 2

ORDINARY DIFFERENTIAL EQUATIONS

2.1   Introduction

The most frequent application of the Laplace Transform is
in determining the solution of ordinary differential equations
with constant coefficients. While this was mentioned in the first
chapter, we now wish to investigate many of the issues which arise
when the Laplace Transform is used in finding the solution.

As we have seen, the effect of taking a Laplace Transform
of an ordinary differential equation having initial conditions is
to transform it into an algebraic equation in the transform function, $F(s)$. Assuming that the algebraic equation can be solved
explicitly, an application of the inversion formula will give the
desired results.

In this chapter we will explore some of the consequences
which follow if we pursue the ideas mentioned above. By beginning
with the definition of a differential equation, we will show how
to determine when a solution exists, a rather crucial question
which should be asked whenever a new problem is studied. We will
then generalize the results to include systems of differential
equations which are the central core of modern applied mathematics.
Once the solution is known to exist, we will proceed to the use of
the Laplace Transform.

While it is absolutely true that transform techniques are

not the only way to solve ordinary differential equations, they,
nevertheless, stand alone as a rich fabric which can be used for
many other applications including partial differential equations,
differential-difference equations and integral equations of the
renewal type. We intend to explore each of these topics in
subsequent chapters.

2.2   Linear Differential Equations With Constant Coefficients

One of the powerful applications of the Laplace Transform,
introduced in Chapter 1, is to find solutions of the Linear,
ordinary differential equations with constant coefficients, such as,

$$(2.2.1) \qquad d^u u/dt^n + a_1 d^{n-1} u/dt^{n-1} + \dots + a_n u = 0$$

when the range of the independent variable goes from zero, or any
constant, to infinity.

Such an equation is known as an n-th order linear
ordinary differential equation, clearly defining the order of the
equation as the order of the highest derivative in the equation,
and having the linear property, namely if $u(t)$ and $v(t)$ are
solutions of (2.2.1), then so is the sum $u(t) + v(t)$.

In addition to specifying the form of the differential
equation (2.2.1), a unique solution, if it exists, requires that
a set of initial conditions also be given. The initial conditions
usually are of the form,

$$(2.2.2) \qquad d^{(n-1)} u(0)/dt^{(n-1)} = c_{n-1}$$

$$u(0) = c_0 \qquad .$$

These are not the only ways to specify the conditions for defining
a solution to (2.2.1).

We assert that the solution of (2.2.1) with the initial
conditions given in (2.2.2), can be solved by Laplace Transform
techniques.

Before we attempt to find a solution to (2.2.1), it is of the utmost importance to examine the conditions under which a continuous solution of (2.2.1) exists and is unique. This is particularly significant when one is trying to find a solution by numerical methods.

We observe that any differential equation of order greater than 1 can be written as a set of first-order differential equations. This may be done by redefining terms as follows. Let

$$u(t) = x_1(t)$$

$$du(t)/dt = x_2(t)$$

$$\cdot$$
$$\cdot$$
$$\cdot$$

$$d^{n-1}u(t)/dt^{n-1} = x_n(t) \quad .$$

Then we can write (2.2.1) as the system

$$dx_1(t)/dt = x_2(t)$$

$$dx_2(t)/dt = x_3(t)$$

$$\cdot$$
(2.2.3)
$$\cdot$$
$$\cdot$$

$$dx_n(t)/dt = -a_1 x_n(t) - a_2 x_{n-1}(t) - \ldots -a_n x_1(t) \quad .$$

By introducing the vector x,

$$(2.2.4) \qquad x(t) = \begin{bmatrix} x_1(t) \\ \cdot \\ \cdot \\ \cdot \\ x_n(t) \end{bmatrix} \quad .$$

The system (2.2.3) becomes

$$(2.2.5) \qquad \frac{d}{dt} \begin{bmatrix} x_1(t) \\ x_2(t) \\ \cdot \\ \cdot \\ \cdot \\ x_n(t) \end{bmatrix} = \begin{bmatrix} 0 & 1 & 0 & \dots & 0 \\ 0 & 0 & 1 & \dots & 0 \\ \cdot & \cdot & \cdot & \dots & \cdot \\ \cdot & \cdot & \cdot & \dots & \cdot \\ \cdot & \cdot & \cdot & \dots & \cdot \\ -a_n & -a_{n-1} & & \dots & -a_1 \end{bmatrix} \begin{bmatrix} x_1(t) \\ x_2(t) \\ \cdot \\ \cdot \\ \cdot \\ x_n(t) \end{bmatrix}$$

or

$$(2.2.6) \qquad dx/dt = Ax$$
$$x(0) = c \quad .$$

To establish the existence of a unique solution to (2.2.1), we shall consider the method of successive approximations. The structure of (2.2.6) is such that we can transform the set of differential equations into a sequence of quadrature problems by observing that (2.2.6) can be expressed in the form,

$$(2.2.7) \qquad dx^{(m)}(t)/dt = Ax^{(m-1)}(t), \quad x^{(m)}(0) = c, \quad m = 1,2, \dots$$

By letting $x^{(0)}(t)$ be an initial guess at the solution, the structure of (2.2.7) allows us to construct a sequence of functions $x^{(m)}(t)$. In fact we can immediately write,

$$(2.2.8) \qquad x^{(m)}(t) = c + \int_0^t A(s)\, x^{(m-1)}(s)\, ds \quad .$$

If the sequence of functions (2.2.8) can be shown to converge to the solution of (2.2.6), then we have proven the existence of such a solution.

To proceed, let us introduce some convenient norms.
Let

$$(2.2.9) \qquad \| x \| = \sum_{i=1}^{n} |x_i|$$

at any time t, where x_1, x_2,...,x_n are the components of the vector x, and

$$(2.2.10) \qquad \| A \| = \sum_{i,j=1}^{n} |a_{ij}| \quad .$$

From (2.2.8) it must be true that

$$(2.2.11) \qquad x^{(m)} - x^{(m-1)} = \int_0^t A(s) \, (x^{(m-1)} - x^{(m-2)}) ds \quad .$$

Taking the norm of both sides, gives

$$(2.2.12) \qquad \| x^{(m)} - x^{(m-1)} \| \leq \int_0^t \| A \| \, \| x^{(m-1)} - x^{(m-2)} \| \, ds \quad .$$

But since

$$\| x^{(1)} - x^{(0)} \| \leq \int_0^t \| A \| \, \| x^{(0)} \| \, ds$$

$$< \| A \| \, \| c \| \, t \quad .$$

Then we have, by induction,

$$(2.2.13) \qquad \| x^{(m)} - x^{(m-1)} \| \leq \| A \|^m \, \| c \| \, |t^m| \, / \, m! \quad .$$

Therefore for any $\varepsilon > 0$, we can find an M for all $t < t_0$, such that for all $m \geq M$,

$$(2.2.14) \qquad \| x^{(m)} - x^{(m-1)} \| < \varepsilon$$

and $x^m(t)$ converges uniformly to a limit. So

$$(2.2.15) \qquad \lim_{m \to \infty} x^{(m)}(t) = x(t) \quad .$$

By virtue of uniform convergence, let $m \to \infty$ on both sides of (2.2.8).

Hence we have,

(2.2.16) $x(t) = c + A\int_0^t x(s)ds, \quad 0 \le t \le t_0$

which is the formal solution of (2.2.6).

Hence we may conclude that the solution of an n-th order ordinary linear differential equation with constant coefficients exists. An easy modification of the foregoing argument yields uniqueness. Knowing this to be true, we can proceed with confidence.

2.3   The Laplace Transform Solution

We begin by considering the solution of linear differential equations with constant coefficients by means of the Laplace Transform. While the results we shall obtain here are in agreement with the solutions obtained in the last sections, we will see a new idea coming into play, namely the transient solution. Consider the first-order scalar equation,

(2.3.1) $du/dt = au + v, \quad u(0) = c$.

Taking the Laplace Transform of both sides, we have,

(2.3.2) $\int_0^\infty e^{-st} (du/dt)dt = a\int_0^\infty e^{-st} udt + \int_0^\infty e^{-st} vdt$.

If we integrate by parts,

(2.3.3) $e^{-st} u\Big|_{0^-}^\infty + s\int_0^\infty e^{-st} udt = a\int_0^\infty e^{-st} udt + \int_0^\infty e^{-st} vdt$.

If we recognize the definition of the Laplace Transform given in Chapter 1, namely,

(2.3.4) $L(u) = \int_0^\infty e^{-st} udt \quad \text{and} \quad L(v) = \int_0^\infty e^{-st} vdt$, then $L(u)$ may be solved for giving the equation,

(2.3.5) $\quad L(u) = c/(s-a) + L(v)/(s-a)$.

The inverse of the first term is ce^{at} while the inverse of the second term is obtained by the convolution theorem given in Chapter 1. The result is

(2.3.6) $\quad u(t) = ce^{at} + \int_0^t e^{a(t-s)} v(s)ds$.

For the vector-matrix case,

(2.3.7) $\quad dx/dt = Ax + y$, $\quad x(0) = c$,

we have,

(2.3.8) $\quad L(x) = (sI - A)^{-1}c + (sI - A)^{-1} L(y)$.

Since the inverse transform of $(sI - A)^{-1}$ is e^{At}, we obtain the expression

(2.3.9) $\quad x = e^{At}c + \int_0^t e^{A(t-s)} y(s)ds$.

We note that in the expression (2.3.9),

$$e^{At} = I + \sum_{n=1}^{\infty} \frac{At^n}{n!}$$.

2.4 Systems of Linear Differential Equations

The Laplace Transform techniques, which we have been considering, can be applied with equal success to systems of linear differential equations with constant coefficients. In this section we shall consider the system.

$$a_{11}du_1/dt + a_{12}u_2 \quad + \ldots + a_{1n}u_n \quad = 0$$
$$a_{21}u_1 \quad + a_{11}du_2/dt + \ldots + a_{2n}u_n \quad = 0$$

(2.4.1)
$$\begin{array}{ccccc} \cdot & & \cdot & \cdot & & \cdot \\ \cdot & & \cdot & \cdot & & \cdot \\ \cdot & & \cdot & \cdot & & \cdot \end{array}$$,

$$a_{n1}u_1 \quad + a_{n2}u_2 \quad + \quad + a_{nn}du_n/dt = 0$$

where the unknown functions $u_1(t)$, $u_2(t),\ldots,u_n(t)$ are subjected to the initial conditions, $u_i(0) = c_i$, $i = 1,2,\ldots,n$.

It can be noted from our previous discussion that systems of higher order differential equations can easily be converted to first-order systems. Consequently we can take (2.4.1) as being a general form of any system of linear differential equations with constant coefficients.

Taking the formal Laplace Transform of system (2.4.1), we obtain the set of transformed algebraic equations,

(2.4.2)
$$\begin{bmatrix} a_{11}s & a_{12} & a_{13} & \cdots & a_{1n} \\ a_{21} & a_{22}s & a_{23} & \cdots & a_{2n} \\ \cdot & \cdot & \cdot & & \cdot \\ \cdot & \cdot & \cdot & & \cdot \\ \cdot & \cdot & \cdot & & \cdot \\ a_{n1} & a_{n2} & a_{n3} & \cdots & a_{nn}s \end{bmatrix} \begin{bmatrix} L(u_1) \\ L(u_2) \\ \cdot \\ \cdot \\ \cdot \\ L(u_n) \end{bmatrix} = \begin{bmatrix} u_1(0) \\ u_2(0) \\ \cdot \\ \cdot \\ \cdot \\ u_n(0) \end{bmatrix}.$$

If we are fortunate enough to be able to explicitly invert the matrix $A(s)$, we have a transformed solution of the form,

(2.4.3)
$$\begin{bmatrix} L(u_1) \\ L(u_2) \\ \cdot \\ \cdot \\ \cdot \\ L(u_n) \end{bmatrix} = A^{-1}(s) \begin{bmatrix} u_1(0) \\ u_2(0) \\ \cdot \\ \cdot \\ \cdot \\ u_n(0) \end{bmatrix}.$$

From (2.4.3) we can construct the formal solution,

$$(2.4.4) \qquad u_j(t) = \frac{1}{2\pi i} \sum_{k=1}^{n} u_k(0) \int_{C_k} e^{st} A_{jk}^{-1}(s) ds$$

where C_k is a contour chosen appropriately for the element $A_{jk}^{-1}(s)$ of the inverse matrix A^{-1}.

In Chapter 6 we discuss the numerical inversion of the Laplace Transform. If such an inversion technique is required to solve (2.4.2) the procedure is to evaluate the matrix $A(s)$ at n values, say $s = 1,2,\ldots,n$. The matrix $A(s)$ is then inverted for each value of s and this information is used with the numerical algorithm described in Chapter 6 to obtain the values of $u(t)$ at a predetermined set of times t_k, $k = 1,2,\ldots,n$. As we shall show later, this numerical procedure can be unstable and should be used with intelligent caution.

2.5   Mismatched Solutions

Let us now consider what appears to be a most elementary and routine computation. We wish to determine the numerical solution to the linear differential equation,

$$(2.5.1) \qquad d^4u/dt^4 - 22d^3u/dt^3 + 39d^2u/dt^2 + 22du/dt - 40u = 0$$

subject to the initial conditions,

$$(2.5.2) \qquad u(0) = 1, \quad du(0)/dt = -1, \quad d^2u(0)/dt^2 = 1,$$

$$d^3u(0)/dt^3 = -1 \quad .$$

This is an ordinary differential equation with prescribed initial values, yet the numerical solution is very unstable. Why?

We can begin to answer this question by examining the roots of the characteristic equation,

(2.5.3) $\lambda^4 - 22\lambda^3 + 39\lambda^2 + 22\lambda - 40 = 0$.

The roots of (2.5.3) occur at -1, 1, 2, 20 thus giving a general solution of (2.5.1) to be

(2.5.4) $u(t) = c_1 e^{-t} + c_2 e^t + c_3 e^{2t} + c_4 e^{20t}$.

The initial conditions were chosen to select as the solution of (2.5.1) and (2.5.2), the function

(2.5.5) $u(t) = e^{-t}$.

If we choose to use a standard numerical integration algorithm to integrate (2.5.1) using (2.5.2), very strange things happen. The solution quickly becomes unstable. How can this surprising phenomenon occur?

The answer is very simple. When we integrate numerically, we introduce small errors in the solution we are actually computing, thus,

(2.5.6) $u(t) = e^{-t} + \varepsilon_1 e^{-t} \varepsilon_2 e^{2t} + \varepsilon_3 e^{20t}$,

where ε_1, ε_2, ε_3 are very small. Yet

(2.5.7) $e^{20} \approx 10^{8.66}$

so that even if $\varepsilon_3 = 10^{-6}$, for moderate values of t, the last term dominates and the required solution is lost in the shuffle. Note carefully the algorithm is calculating a solution to (2.5.1) but it is not getting the correct one corresponding to (2.5.2).

This phenomenon occurs whenever the characteristic roots vary greatly in magnitude. Of course, if we want the solution corresponding to the largest root, we have no problem.

However, by using the Laplace Transform we can overcome or

circumvent the difficulty in this case. Let us write Eq. (2.5.1) as a first-order system

$$(2.5.8) \qquad u_1'(t) = u_2(t) \qquad\qquad u_1(0) = 1$$

$$u_2'(t) = u_3(t) \qquad\qquad u_2(0) = -1$$

$$u_3'(t) = u_4(t) \qquad\qquad u_3(0) = 1$$

$$u_4'(t) = 22u_4(t) - 39u_3(t) \qquad u_4(0) = -1 \quad .$$
$$- 22u_2(t) + 40u_1(t)$$

Taking the Laplace Transform of (2.5.8) and setting $w_i = L(u_i)$,

$$(2.5.9) \qquad sw_1 = w_2 + 1$$

$$sw_2 = w_3 - 1$$

$$sw_3 = w_4 + 1$$

$$sw_4 = 22w_4 - 39w_3 - 22w_2 + 40w_1 + 1 \quad .$$

If we solved this linear system explicitly, we would find, of course, that

$$(2.5.10) \qquad w_1 = \frac{P_1(s)}{P_2(s)} = \frac{1}{s+1} \quad .$$

In general, however, when this equation arises as part of a comprehensive mathematical formulation we expect to solve (2.5.9) numerically for $s = 1,2,\ldots,N$ using any number of standard algorithms for solving linear systems of algebraic equations.

This direct procedure leads to difficulty in this case since $P_2(s)$ vanishes at $s = 1$ and $s = 2$. These zeros are cancelled analytically by corresponding zeros in the numerator, but not necessarily numerically. To avoid this pitfall, we use an alternative set of s-values, namely $s = 3/2, 5/2, 7/2, R + 1/2$. With this information we can invert the Laplace Transform numerically by the methods described in Chapter 6.

2.6   Behavior Of The Higher Derivatives

An interesting problem is the following. Given the linear differential equation

$$(2.6.1) \qquad x''(t) + ax'(t) + bx(t) = g(t)$$
$$x(0) = x_0$$
$$x'(0) = x_0' \qquad ,$$

we ask, what is the behavior of $x'(t)$?

Of course, we could solve (2.6.1) analytically and differentiate. Also we could solve (2.6.1) numerically and then proceed to differentiate the results numerically but, as we should now suspect, this procedure can lead to serious numerical difficulties.

An alternative is to use the Laplace Transform. Rewriting (2.6.1) as a first-order linear system,

$$(2.6.2) \qquad u_1' = u_2 \qquad\qquad u_1(0) = x_0$$
$$u_2' = g(t) - au_2 - bu_1 \qquad u_2(0) = x_0'$$

and taking Laplace Transforms, setting $w_i = L(u_i)$, we have,

$$(2.6.3) \qquad sw_1 = w_2 + x_0$$
$$sw_2 = G(s) - aw_2 - bw_1 + x_0'$$

where $G = L(g)$ is known.

To solve our problem, we need only solve (2.6.3) for w_2 and invert the expression. If we carry out the algebra, we get,

$$(2.6.4) \qquad w_2(s) = \frac{sG(s) + sx_0{}' - bx_0}{s^2 + as + b} \, .$$

Depending on the precise form of $G(s)$, we can invert (2.6.4) analytically or by numerical methods given in Chapter 6. In either case, we obtain the behavior of $\dot{x}(t)$ without having to solve for $x(t)$.

2.7 An Example

The Laplace Transform is far from being merely a mathematical nicety for it has many applications in the field of engineering. We shall study one such problem occurring in electrical engineering. In this example we wish to know what happens when a rectangular voltage pulse is applied to a series RC circuit. The circuit itself is shown in Fig. 2.1 together with a representation of the applied pressure pulse.

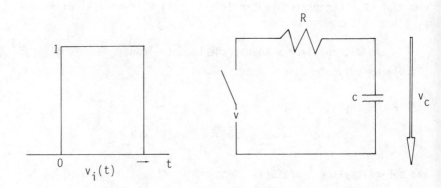

Fig. 2.1

Precisely the problem is as follows. A rectangular voltage pulse of unit height and duration T is applied to a series RC circuit at $t = 0$. Determine the voltage across the capacitance as a function of time.

The voltage pulse can be represented mathematically as a difference of two step functions, for from the definition,

$$(2.7.1) \qquad U(t) = \begin{cases} 0 & t \le 0 \\ 1 & t > 0 \end{cases} \quad ,$$

we can immediately define

$$(2.7.2) \qquad v_i(t) = U(t) - U(t-T) \quad .$$

The governing equation for this system is derived from the basic ideas of electrical engineering. In electrical units, the desired response is $v_c = q/c$, where q is the electrical charge, c is the capacitance, v_c is the voltage across the capacitor, and R is the electrical resistance.

Since for current i,

$$(2.7.3) \qquad i = dq/dt = cdv_c/dt \quad ,$$

the governing differential equation for the system is

$$(2.7.4) \qquad RCdv_c/dt + v_c = v_i(t) = U(t) - U(t-T) \quad .$$

Let $V_c(s) = L(v_c(t))$. Taking the Laplace Transform of (2.7.4)

$$(2.7.5) \qquad RC(sV_c(s) - v_c(0+)) + V_c(s) = L(U(t) - U(t-T)) \quad ,$$

Now,

$$L(U(s)) = \int_0^\infty e^{-st} U(t)dt = \int_0^\infty e^{-st} dt = 1/s \quad .$$

Equation (2.7.5) then becomes,

$$(2.7.6) \qquad RC(sV_c(s) - v_c(0+)) + V_c(s) = (1 - e^{-sT})/s$$

and solving for $V_c(s)$, we get, setting $v_c(0+) = 0$,

(2.7.7) $\qquad V_c(s) = (1 - e^{-sT})/(s(RCs + 1))$.

We see from examining the function $V_c(s)$ that it is an analytic function in the complex s plane except for two poles, one at $s = 0$, the other on the negative real axis at $s = -1/RC$. The final result, $v_c(t)$ is found by taking the inverse Laplace Transform of (2.7.7), that is,

(2.7.8) $\qquad v_c(t) = \dfrac{1}{2\pi i} \displaystyle\int_{-\infty i+a}^{\infty i+a} e^{st}(1 - e^{-sT})/(s(RCs + 1))ds$

$\qquad\qquad\quad = \dfrac{1}{2\pi i} \displaystyle\int_{-\infty i+a}^{\infty i+a} (e^{st} - e^{s(t-T)})/(s(RCs + 1))ds$.

If we consider the term,

(2.7.9) $\qquad 1/(s(RCs + 1)) = (1/RC)(RC/s - RC/(s + 1/RC))$

$\qquad\qquad\qquad\qquad\quad = (1/s - 1/(s + 1/RC))$.

Substituting (2.7.9) into (2.7.8),

(2.7.10) $\qquad v_c(t) = \dfrac{1}{2\pi i} \displaystyle\int_{-\infty i+a}^{\infty i+a} (e^{st} - e^{s(t-T)})/(1/s - 1/(s + 1/RC))ds$

Rewriting (2.7.10),

(2.7.11) $\qquad v_c(t) = \dfrac{1}{2\pi i} \displaystyle\int_{-\infty i+a}^{\infty i+a} e^{st}/s - e^{st}/(s + 1/RC)$

$\qquad\qquad\qquad\quad - \dfrac{e^{s(t-T)}}{s} + e^{-sT}e^{st}/(s + 1/RC)\Big)ds$

$\qquad\qquad\quad = L^{-1}(1/s) - L^{-1}(e^{-sT}/s) - L^{-1}(1/(s + 1/RC))$

$\qquad\qquad\qquad + L^{-1}(e^{-sT}/(s + 1/RC))$.

It is left as an exercise for the reader to show that,

$\qquad L^{-1}(1/s) = U(t)$

$$L^{-1}(e^{-sT}/s) = U(t-T)$$

$$L^{-1}(1/(s+1/RC)) = e^{-t/RC} U(t)$$

$$L^{-1}(e^{-sT}/(s+1/RC)) = e^{-(t-T)/RC} U(t-T) \quad .$$

Substituting into (2.7.11) give us the solution,

$$(2.7.12) \quad v_c(t) = (1 - e^{-t/RC}) U(t) - (1 - e^{-(t-T)/RC}) U(t-T) \quad .$$

We see immediately that the model predicts that there is no response across the capacitor for $t < 0$ and in fact for $t = 0$. Neither does the model predict any oscillations in the solution. In fact the solution is shown in Fig. 2.2.

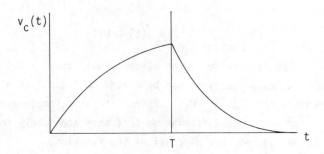

Fig. 2.2

As one would perhaps expect, there is a discontinuity in slope at $t = 0$ and $t = T$. The existence of a set of discontinuities in slope leads one to question if such a discontinuity is reflected in the basic differential equation governing the system. Taking the first derivative of (2.7.12),

$$(2.7.13) \quad dv_c/dt = (1 - e^{-t/RC}) \delta(t) - (1 - e^{-(t-T)/RC}) \delta(t-T)$$

$$+ (e^{-t/RC}/RC)U(t) - (e^{T/RC} e^{-t/RC}/RC)U(t-T)$$

and rearranging terms,

$$(2.7.14) \quad RCdv_c/dt = RC(1 - e^{-t/RC})\delta(t)$$

$$- RC(1 - e^{-(t-T)/RC})\delta(t-T)$$

$$+ (-1 + e^{-t/RC})U(t)$$

$$- (-1 + e^{-(t-T)/RC})U(t-T)$$

$$+ U(t) - U(t-T) \quad .$$

Or,

$$(2.7.15) \quad RCdv_c/dt + v_c = RC(1 - e^{-t/RC})\delta(t)$$

$$- RC(1 - e^{-(t-T)/RC})\delta(t-T)$$

$$+ U(t) - U(t-T) \quad .$$

The appearance of the discontinuity in (2.7.15) in the form of a delta function, can be traced to the differentiation of the unit step function $U(t)$. Since it is an indeterminate form as it now stands we must examine this term separately to determine its actual value. Let us consider the function,

$$(2.7.16) \quad g(t) = (1 - e^{-t/RC})\delta(t) - (1 - e^{-(t-T)/RC})\delta(t-T) \quad .$$

Since the discontinuities occur at different times, they can be treated separately. If we write the delta function as the following limit,

$$(2.7.17) \quad \delta(t) = \lim_{a \to 0} ((U(t) - U(t-a))/a)$$

then the first term in (2.7.16) can be written as,

$$(2.7.18) \quad g_1(t) = (1 - e^{-t/RC})(\lim_{a \to 0} ((U(t) - U(t-a))/a) \quad .$$

Since a is independent of t, the limit can be taken out consequently and we can write,

$$(2.7.19) \qquad g_1(t) = \lim_{a \to 0} \left[\left((1 - e^{-t/RC})/a \right) (U(t) - U(t-a)) \right]$$

If we let $t \to 0$ in the same order as a, i.e., let $a = kt$, then

$$(2.7.20) \qquad g_1(t) = \lim_{t \to 0} \left[\left((1 - e^{-t/RC})/kt \right) (U(t) - U(t-kt)) \right] \quad .$$

By expanding the exponential in a power series and letting $t \to 0$, the first term approached a constant, $1/kRC$, while the second term approaches zero. Therefore $g_1(t) = 0$ at $t = 0$, the same argument holds at $t = T$. Hence we can say positively that $g_1(t)$ is zero for all t and the solution (2.7.12) is the correct one.

Problems

1. Show that the Laplace Transform can be used to reduce the solution of $u'' + (a_0 + a_1 t)u = 0$ to that of a first-order differential equation.

    Does this have any advantage from a numerical standpoint?

2. Obtain the corresponding results for the equation

    $$u^{(n)} + (a_0 + a_1 t)u^{(n-1)} + \ldots + (a_{2n-2} + a_{2n-1} t)u = 0 \quad .$$

    These are examples of differential equations with variable coefficients. Can you construct other examples where the Laplace Transform technique can be useful?

3. Solve the equation

    $$du/dt + au = b \int_0^1 u \, dt \, , \qquad u(0) = c \quad .$$

4. Solve the equation

$$du/dt + au = \phi(\int_0^1 udt), \qquad u(0) = c \quad .$$

5. Study the solutions of the equation

$$u''(t) + (1 + ae^{bt})u = 0$$

by taking Laplace Transforms and considering the resulting difference equation.

6. Use the same technique to study the Mathieu equation

$$u''(t) + (1 + a \cos bt)u = 0 \quad .$$

7. Consider the differential equation,

$$(1) \qquad dx/dt = x, \qquad x(0) = 1 \quad .$$

Using the method of successive approximation, write the sequence of quadrature problems generated by (1). Write down the quadrature solution. Let $x^{(0)}(t) = 1$, compute the first 5 successive approximations. If $\varepsilon = 10^{-6}$, and consider a measure of error to be

$$|x^{(5)}(t) - x^{(4)}(t)| < \varepsilon \quad ,$$

find the range of t for which the inequality holds. Generalize the result to the case where

$$|x^{(n+1)}(t) - x^{(n)}(t)| < \varepsilon \quad .$$

8. Consider the equation for damped oscillations

$$d^2x/dt^2 + a \, dx/dt + bx = 0$$

$$x(0) = x_0$$

$$(2) \qquad dx(0)/dt = x_0' \quad .$$

By finding the roots of the characteristic equation, write

the general solution of (2). Using the Laplace Transform
technique, find the solution of (2). The task of the applied
mathematician is often to model an existing physical
phenomenon by a mathematical expression of some sort. Unless
the model is extremely accurate, the system constants cannot
be accurately computed. In these cases, intuitive feeling
about the behavior of the system can be very useful. For
example, if the system is known to go to zero for large
times, then what can be said about the location of the roots
of the characteristic equation in the complex plane? If the
system reaches a steady state constant value for large t,
where are the roots located now? Do the initial conditions
determine the steady state value? What is the precise
effect? Critical damping of the system occurs when the
steady state value is reached without oscillations. From
the Laplace Transform solution, show what relation must exist
between the system parameters and the initial conditions for
this to occur.

9. Consider the system of differential equations

$$du/dt + a_{11}u_1 + a_{12}u_2 = 0 , \qquad u_1(0) = b_1$$

$$du_2/dt + a_{21}u_1 + a_{22}u_2 = 0 , \qquad u_2(0) = b_2 .$$

By explicitly inverting a 2×2 matrix, show that the Laplace
Transform function F(s) is,

$$F(s) = \frac{1}{(s+a_{11})(s+a_{22}) - a_{12}a_{21}} \begin{bmatrix} s+a_{22} & -a_{12} \\ -a_{21} & s+a_{11} \end{bmatrix} \begin{bmatrix} b_1 \\ b_2 \end{bmatrix} .$$

Hence show,

$$u_1(t) = \frac{1}{2\pi i}\left\{ b_1 \int_{C_1} \frac{e^{st}(s+a_{22})ds}{(s-s_1)(s-s_2)} - a_{12}b_2 \int_{C_2} \frac{e^{st}ds}{(s-s_1)(s-s_2)} \right\}$$

$$u_2(t) = \frac{1}{2\pi i}\left\{ -a_{21}b_1 \int_{C_3} \frac{e^{st}ds}{(s-s_1)(s-s_2)} + b_2 \int_{C_4} \frac{e^{st}(s+a_{11})ds}{(s-s_1)(s-s_2)} \right\}$$

where s_1, s_2 are the roots of the characteristic equation and $C_1,\ldots,C_4$ are appropriately chosen contours. Show that if $a_{11}a_{22} = a_{12}a_{21}$, we would expect a steady state solution when $a_{11} + a_{22} > 0$.

Discuss the locations of the contours in the complex s-plane.

Chapter 3

DIFFERENTIAL-DIFFERENCE EQUATIONS

3.1   Introduction

While the Laplace Transform technique has been successfully applied to the solution of ordinary differential equations, as we have shown in the last chapter, a different class of differential equations also have a structure suggestive of solution by means of Laplace Transforms. These are the differential-difference equations which appear quite frequently in problems in biomechanics. Because they present a clear application of the Laplace Transform, we shall devote this chapter to a study of this special form of ordinary differential equations.

By a differential-difference equation we shall, in general, mean an equation of an unknown function and its derivatives, evaluated at arguments which differ by any of a fixed number of values. Examples of such equations are

$$(3.1.1) \qquad u''(t) - u'(t-1) + u(t) = 0 \quad ,$$

$$(3.1.2) \qquad u'(t) - u(t-1) - u(t-\sqrt{2}) = 0 \quad ,$$

$$(3.1.3) \qquad u'(t) - 2u(t) + u'(t-1) - 2u(t-1) = e^{2t} \quad .$$

The equations are characterized by the property that the derivative and the functional values are no longer evaluated at the same time t, but the equation defines a relationship between properties of the solution at different values of the independent variable. Thus, for example, in (3.1.1) the equation defines a relationship between the second derivative of u at the current time t, the first derivative at a time t minus one unit and the function u at t.

In this chapter we shall restrict our study to problems in which u will be regarded as a function of a single independent variable t so that all derivatives will appear as ordinary rather than partial derivatives. The differential-difference equation will be classified by defining its orders. Thus by the differential order of the equation we mean the highest derivative appearing in the equation and the difference order of the equation as one less than the number of distinct arguments appearing. For example, Eq. (3.1.1) is of differential order 2 and of difference order 1. Note that $u(t) = u(t-0)$, zero being one argument.

We will further restrict ourselves to linear differential equations with at most time varying coefficients of the form,

$$(3.1.4) \qquad \sum_{i=0}^{m} \sum_{j=0}^{n} a_{ij}(t) u^{(j)}(t-s_i) = f(t) \qquad ,$$

where m, n are positive integers, where $0 < s_0 < s_1 < \cdots < s_m$ and where $f(t)$ and the $(n+1)(m+1)$ functions $a_{ij}(t)$ are defined on some interval of the real t-axis. Since $u^{(j)}(t)$ is the notation of the j-th derivative of $u(t)$ with respect to t, we can easily see that Eq. (3.1.4) can be characterized as being of the n-th differential order and the (m-1)st difference order. Further, linear differential-difference equations with constant coefficients have the obvious form,

$$(3.1.5) \qquad \sum_{i=0}^{m} \sum_{j=0}^{n} a_{ij} u^{(j)}(t-s_i) = f(t) \qquad .$$

By now it would seem rather clear that differential-difference equations closely resemble ordinary differential equations. In fact, in the limit, when the set of arguments (s_i) approach zero, the differential-difference equation becomes the standard differential equation. Because of the similarity in structure between the two types of equations, we would expect some degree of similarity in the solutions of the equations. Here we must be very careful, for the characterization of solution of differential-difference equations is much more complicated.

To understand the behavior of differential-difference equations, we must, nevertheless, ask questions which are very similar to the ones which have been asked concerning ordinary linear differential equations. Among these are the following:

(1) What is the correct formulation of the initial values problem?

(2) Can the solution be represented by sums of particular solutions?

(3) What is the asymptotic behavior of the solution?

(4) What can be said about the concept of stability of solutions?

3.2   Examples

The notion of a differential-difference equation can best be introduced thru the careful examination of the few specific examples. The basic ideas we wish to illustrate here is the form of the equation, the method for starting a solution to such an equation and how the general solution is constructed from the initial information given about the equation.

An interesting example is the differential-difference equation

$$(3.2.1) \qquad u'(t) = u(t-1) \qquad .$$

This equation is first-order in both the difference and differential indices. We seek a solution of (3.2.1) which is continuous for all $t > 0$ and is a solution of (3.2.1) for all $t > 1$. A close look at Eq. (3.2.1) reveals it as essentially a definition of the slope of the function $u(t)$ evaluated at t as a function of u at a time t shifted back by one unit. Hence, it is easy to see that $u(t)$ can be set to an arbitrary continuous function over the initial interval of length one. Once this has been done, however, the solution $u(t)$ is uniquely determined from (3.2.1) for all $t > 0$. We should point out that in the case of a differential-difference equation the "initial condition" is an *initial interval* over which a continuous function must be specified. For the ordinary differential equation, the initial interval has been replaced by an initial condition at a single point.

To construct a solution to (3.2.1), we shall, for example, let $u(t) = 1$ for $t < 1$. Then if (3.2.1) is to hold for $t > 1$, the values of $u'(t)$ are determined for $1 \le t \le 2$. In fact we have,

$$u(t) = t = 1 + (t - 1), \qquad 1 \le t \le 2 \quad .$$

Since $u(t)$ is now known for $1 \le t \le 2$, Eq. (3.2.1) determines $u(t)$ for $2 \le t \le 3$. In fact,

$$u(t) = 1 + (t - 1) + (t - 2)^2/2 \quad , \qquad 2 \le t \le 3 \quad .$$

We can proceed in this fashion as far as we please, extending the definition of $u(t)$ from one interval to the next and we find, by induction, the relation,

$$(3.2.2) \qquad u(t) = \sum_{j=0}^{N} (t - j)^j/j!, \quad N \le t \le N+1, \quad N = 0,1,2, \ldots$$

The above example illustrates one of the fundamental method available for the discussion of differential-difference equations, the methods of continuation by which the solution is extended forward, i.e., in the direction of increasing t, from interval to interval. This method provides a means of proving that an equation

has a solution and moreover gives a procedure for the actual
calculation of the solution. The example given also illustrates
the fact that a differential-difference equation ordinarily has a
great variety of solutions, one of which can be singled out by
requiring that it has specified values over a certain t-interval,
just as one solution of a first-order differential equation can be
singled out by requiring that it has a specific value at a certain
point. Such additional conditions on a solution are called
boundary conditions. From the above remarks, we see that a
sensible boundary condition for Eq. (3.2.1) is the condition

$$(3.2.3) \qquad u(t) = g(t), \qquad 0 < t < 1 \qquad ,$$

when $g(t)$ is any preassigned, real, continuous function.

A boundary condition of the type (3.2.3) which prescribes
the solution $u(t)$ in an initial interval of values of t, from
which a solution can be continued, is also sometimes called an
initial condition. Of course, it would be possible to impose
other kinds of boundary conditions on the solutions of (3.2.1) and
in fact certain others occur naturally in various applied problems.
It is of fundamental importance for the mathematician studying
differential equations, differential-difference equations or
other functional equations, to determine boundary conditions of
various sorts that are of the correct severity to permit the
existence of a unique solution of a specific type.

A somewhat different situation is illustrated by the
equation,

$$(3.2.4) \qquad u'(t) = u(t-1) + 2u'(t-1) \qquad .$$

Again let us suppose that $u(t) = 1$ for $0 < t < 1$ and require that
$u(t)$ be continuous for $t > 0$. If (3.2.4) is to be satisfied for
$t > 1$, we must have $u'(t) = 1$ for $1 < t < 2$ and therefore

$$u(t) = t, \qquad 1 < t < 2 \qquad .$$

If (3.2.4) is to be satisfied for $2 < t < 3$, we must, therefore, have $u'(t) = t + 1$, $2 < t < 3$, and

$$u(t) = \frac{1}{2} t^2 + t - 2, \qquad 2 < t < 3 \quad .$$

By continuing this process we can continue $u(t)$ as far forward as we please. Moreover, the solution $u(t)$ obtained has a derivative discontinuity at every positive integer value of t. The equation is satisfied at each such value only in the sense of left-hand limits and right-hand limits. In fact, though there are exceptions, it is generally not true that a solution of (3.2.4) will be continuous and have continuous first derivatives for all $t > 0$.

As a final example, let us consider the equation,

(3.2.5) $u'(t - 1) = u(t)$,

again subject to the initial conditions of the form in (3.2.3). It is not hard to see that this solution can be continued backwards by the same process as was used in (3.2.1). If, however, we attempt to continue the solution forward we have,

$$u(t) = g'(t - 1), \qquad 1 < t < 2 \quad ,$$

provided g is differentiable. This determines $u(t)$ for $1 < t < 2$. Provided $g(t)$ is twice differentiable for $0 < t < 1$, $u(t)$ is differentiable for $1 < t < 2$, and Eq. (3.2.5) can be used to define $u(t)$ for $2 < t < 3$. We see that this continuation process yields a solution for all $t > 0$ only if the initial function $g(t)$ possesses derivatives of all orders for $0 < t < 1$.

3.3   Types of Differential-Difference Equations

In establishing the ideas which are central to the notion of differential-difference equations, it is useful and often necessary to categorize the different forms that the equation can take. As we have seen before, the general form of a linear differential-difference equation with constant coefficients can be written,

$$(3.3.1) \qquad \sum_{i=0}^{m} \sum_{j=0}^{n} a_{ij} u^{(j)}(t - c_i) = f(t) \quad ,$$

which describes a differential-difference equation of differential order n and difference order $(m-1)$.

A reduced example of (3.3.1) is given by

$$(3.3.2) \qquad a_0 u'(t) + a_1 u'(t-c) + b_0 u(t) + b_1 u(t-c) = f(t) \quad ,$$

which is of order 1 in derivatives and differences. It is desirable to further classify equations of the form (3.3.2) in several categories. An equation in the form (3.2.2) is said to be of retarded type if $a_0 \neq 0$ and $a_1 = 0$. It is said to be of neutral type if $a_0 \neq 0$ and $a_1 \neq 0$. It is said to be of advanced type if $a_0 = 0$ and $a_1 \neq 0$. If $a_0 = a_1 = 0$, the equation is a pure difference equation, a type of functional equation which has been treated in great detail. If $b_0 = b_1 = 0$, it reduces to a pure difference equation. If $a_0 = b_0 = 0$ or $a_1 = b_1 = 0$, it is an ordinary differential equation.

In some applications, in which t is time, an equation of retarded type may represent the behavior of a system in which the rate of change of the quantity under investigation depends on past and present history. An equation of neutral type may represent a system in which the present rate of change of the quantity depends on the past rate of change as well as the past and present values of the quantity. An equation of the advanced type may represent a system in which the rate of change of a quantity depends on present and future values of the quantity (or alternatively, in which the present value of the quantity depends on the past value and the past rate of change).

Since t usually represents the time in applications, we shall ordinarily be interested in continuing a solution in the direction of increasing time t. One should note, however, that the substitution $t' = -t$ converts an equation of the retarded type into an equation of the advanced type in t' and vice versa, and converts an equation of neutral type to another equation of neutral type. Thus, without loss of generality, confine our investigations to increasing values of t.

3.4   Existence and Uniqueness

To the numerical analyst, and the applied mathematician, it is very important to be able to determine whether a sought for solution to a differential-difference equation exists. Aside from the obvious embarrassment of attempting to solve numerically for a nonexistent solution, the proven existence of a solution to a given equation gives some authority to the underlying mathematical model which is being formulated. We believe it is important to lay before the reader an informal argument of the existence of solutions to differential-difference equations so that the results given in the remaining part of this chapter will be on a firm foundation.

We shall begin our discussion by considering the existence and uniqueness of solutions of the equations of retarded type,

$$(3.4.1) \qquad a_0 u'(t) + b_0 u(t) + b_1 u(t-c) = f(t)$$

subject to the initial condition of the form $u(t) = g(t)$ for $t_0 < t < t_0 + c$. We first observe that the translation $t - t_0 = t'$ converts the equation into an equation of the same form subject to initial conditions over $0 < t < c$. Therefore, without loss of generality, we shall take as initial conditions,

$$(3.4.2) \qquad u(t) = g(t), \qquad 0 \le t \le c \qquad .$$

In order to describe more easily the conditions imposed on a solution of (3.4.1), we shall introduce some standard notations. The set of all real functions, f having k continuous derivatives in an open interval, $t_1 < t < t_2$ is denoted by $C^k(t_1,t_2)$. It is also convenient to extend this definition to intervals which are not open. As usual the notations $[t_1,t_2]$, $[t_1,t_2)$, $(t_1,t_2]$ and (t_1,t_2) denote respectively the intervals $t_1 \le t \le t_2$, $t_1 \le t < t_2$, $t_1 < t \le t_2$ and $t_1 < t < t_2$. To account for the closed intervals, we say a function $f(t)$ is of class C^k on $[t_1,t_2)$ if it is of class C^k on (t_1,t_2), if it has a right-hand k-th derivative at t_1 and if the function $f^{(k)}(t)$ defined over $t_1 < t < t_2$ by those values is continuous from the right at t_1. A function $f(t)$ is said to be of class C^k on $(t_1,t_2]$ if these statements are valid when "right" is replaced by "left" and "t_1" by "t_2". If both these conditions hold, $f(t)$ is said to be of class C^k on $[t_1,t_2]$.

The general result of this section is embodied in the following. Suppose that $f(t)$ is of class C^1 on $[0,\infty)$ and that $g(t)$ is of class C^0 on $[0,c]$. Then there exists one and only one function $u(t)$ for $t \ge 0$ which is continuous for $t \ge 0$, which satisfies (3.4.2) and which satisfies (3.4.1) for $t > c$. Moreover this function $u(t)$ is of class C^1 on (c,∞) and of class C^2 on $(2c,\infty)$. If $g(t)$ is of class C^1 on $[0,c]$, $u'(t)$ is continuous at $t=c$, if and only if,

$$(3.4.3) \qquad a_0 g'(c-0) + b_0 g(c) + b_1 g(0) = f(c) \qquad .$$

If $g(t)$ is of class C^2 on $[0,c]$, $u''(t)$ is continuous at 2c, if either (3.4.3) holds or else $b_1 = 0$ and only in these cases.

To prove this result, let

$$v(t) = f(t) - b_1 u(t-c) \qquad ,$$

then (3.4.1) can be written as

$$a_0 u'(t) + b_0 u(t) = v(t)$$

or by multiplying both sides by $e^{b_0 t/a_0}$, and rearranging terms, we have,

(3.4.4) $(d/dt)(a_0 u(t) \exp(b_0 t/a_0)) = v(t) \exp(b_0 t/a_0)$.

We argue that since $f(t)$ is continuous with its first derivative for $t > 0$, and $g(t)$ is continuous on $[0,c]$, then $v(t)$ must be continuous on $[0,c]$. By integrating (3.4.4), we construct a unique function $u(t)$ which satisfies (3.4.1) for $c \le t \le 2c$, and for which $u(c) = g(c)$. Since this function is continuous, $v(t)$ is of class C^0 on $[c,3c]$. From (3.4.4) it follows that there is a unique continuous function $u(t)$ which satisfies (3.4.1) for $c < t < 3c$. Since, clearly, this process can be repeated as often as we please, we have established the existence and uniqueness for the function $u(t)$.

From (3.4.1) we have,

(3.4.5) $a_0 u'(t) = f(t) - b_0 u(t) - b_1(t-c)$, $t > c$.

Since we have already shown $u(t)$ is $C^{(0)}$ on $[0,\infty)$, it follows that $u'(t)$ is C^0 on (c,∞). Moreover the right-hand member of Eq. (3.4.5) is differentiable, and, in fact,

(3.4.6) $a_0 u''(t) = f'(t) - b_0 u'(t) - b_1 u'(t-c)$, $t > 2c$.

The right-hand member of Eq. (3.4.6) is of class C^0 on $(2c,\infty)$ and therefore $u(t)$ is of class C^2 on $(2c,\infty)$.

If $g(t)$ is of class C^1 on $[0,c]$, then by (3.4.1), when $t = c$,

$$a_0 u'(c+0) = f(c) - b_0 g(c) - b_1 g(0)$$

whereas since $g(t)$ has a continuous first derivative at $t = c$,

$$a_0 u'(c-0) = a_0 g'(c-0) \quad .$$

Therefore, $u'(t)$ is continuous at c, if and only if,

$$(3.4.7) \qquad a_0 g'(c-0) + b_0 g(c) + b_1 g(0) = f(c) \quad .$$

If $g(t)$ is of class C^2 on $[0,c]$, we see from (3.4.6), which holds also for $0 < t < 2t$, that $u''(t)$ is continuous at $2c$, if and only if,

$$b_1(u'(c+0) - u'(c-0)) = 0 \quad .$$

This will be the case if and only if either u' is continuous at c or $b_1 = 0$. In the latter case, (3.4.1) is a pure differential equation.

Finally, it can be shown, without loss of generality, that if $g(t)$ is of class C^2 on $[0,c]$ and (3.4.7) holds, then $u(t)$ is of class C^2 on $[0,\infty)$. Thus if the initial function is continuous with continuous first derivatives on the closed interval, the solution of (3.4.1) exists, and is unique with continuous first and second derivatives on the interval $(0,\infty)$.

3.5   Exponential Solutions

In the last section we described the continuation process whereby the solution of a differential-difference equation can be extended from one interval to the next. It has been shown that if the solution exists, it is continuous and is unique. Rarely can the continuation solution be written in a closed form and even when this is possible, such a formula may not be particularly useful in looking for certain properties of the solution. One such of particular significance is its behavior as t becomes infinitely large. There is a second method for building the solution to a differential-difference equation which is very useful. This second fundamental method is considered in this

section. The results obtained here will lead directly to the use
of Laplace Transforms in the solutions of differential-difference
equations.

Let us define the linear operator, $L(u)$,

$(3.5.1)$ $L(u) = a_0 u'(t) + b_0 u(t) + b_1 u(t - c)$.

We first observe that one property of the linear operator is

$$L(c_1 u_1 + c_2 u_2) = c_1 L(u_1) + c_2 L(u_2)$$.

The differential-difference equation can be written in operator
notation as,

$(3.5.2)$ $L(u) = f$,

and as a consequence of linearity, we observe that if $v(t)$ is a
solution of $L(u) = f$, and if $w(t)$ is a solution of $L(u) = 0$,
then $v + w$ is a solution of $L(u) = f$. This is clear, since
$L(v + w) = L(v) + L(w) = f + 0 = f$. Using this result we can see
that if $L(u) = f$, subject to the initial condition that $u = g$
on $t_0 < t < t_0 + c$ can be obtained by adding the solutions v and
w of two simpler problems, namely; the solution v of $L(v) = 0$,
$v = g$ on $t_0 < t < t_0 + c$, and the solution w of $L(w) = f$, $w = 0$
on $t_0 < t < t_0 + c$.

Let us consider the problem of finding a solution of
$L(u) = 0$. If we set $u(t) = e^{st}$, where s is an unknown parameter, then, because of the property of exponentials, namely
$e^{(a + b)} = e^a e^b$, then $L(u) = 0$ can be written as,

$(3.5.3)$ $L(e^{st}) = (a_0 s + b_0 + b_1 e^{-cs}) e^{st} = 0$.

This equation is satisfied only if,

$(3.5.4)$ $h(s) = (a_0 s + b_0 + b_1 e^{-cs}) = 0$.

We have, by this substitution, succeeded in transforming the homogeneous differential-difference equation, $L(u) = 0$, into a transcendental equation, $h(s) = 0$, which is defined as the characteristic function of $L(u)$. Furthermore, exponential solutions of the equation $L(u) = 0$ exist only at the roots of the characteristic equation, $h(s) = 0$. These values of s are known as the characteristic roots of L.

Corresponding to each characteristic root there is a solution (which may be complex) of $L(u) = 0$ and to distinct roots correspond linearly independent solutions. Moreover, a multiple root gives rise to several independent solutions as we will now show.

We first observe that,

$$(3.5.5) \qquad h'(s) = a_0 - b_1 c e^{-cs}$$

and by further differentiation,

$$(3.5.6) \qquad h^{(k)}(s) = (-1)^k b_1 c^k e^{-cs} \ , \qquad k = 2, 3, \ldots$$

For any $n \geq 1$, we have using (3.5.1)

$$(3.5.7) \qquad L(t^n e^{st}) = a_0(t^n s e^{st} + n t^{n-1} e^{st}) + b_0 t^n e^{st}$$
$$+ b_1 (t - c)^n e^{s(t-c)} \ .$$

If $(t - c)^n$ is expanded by the binomial theorem, we see that the coefficients of $t^{n-k} e^{st}$ $(0 < k < n)$ in (3.5.7) is,

$$(3.5.8) \qquad \binom{n}{k} h^{(k)}(s) \ .$$

Consequently,

$$(3.5.9) \qquad L(t^n e^{st}) = e^{st} \sum_{k=0}^{n} \binom{n}{k} t^{n-k} h^{(k)}(s) \ .$$

From this equation we see that $L(t^n e^{st}) = 0$ for any integer n in the range $0 < n < m-1$ if s is a characteristic root of multiplicity m since $h(s), h'(s), \ldots, h^{(m-1)}(s)$ must all vanish for this value of s. Thus, a root s of multiplicity m gives rise to m functions $e^{st}, te^{st}, \ldots, t^{m-1}e^{st}$ which are solutions of $L(u) = 0$ for all real t. As is well known, these m functions are linearly independent over any interval. Since the equation $L(u) = 0$ is linear and homogeneous, $p(t)e^{st}$ is evidently a solution if $p(t)$ is any polynomial of degree not greater than $m-1$. Hence we can say that the equation

(3.5.10) $L(u) = a_0 u'(t) + b_0 u(t) + b_1 u(t - c) = 0$

is satisfied by

$$\sum p_r(t) e^{s_r t} \quad ,$$

where s_r is any sequence of characteristic roots of L, $p_r(t)$ is a polynomial of degree less than the multiplicty of s_r and the sum is either finite or is infinite with suitable conditions to insure convergence.

Although the results of this section are similar to those in the theory of ordinary differential equations, there is one very important difference. There are, in general, infinitely many characteristic roots (and therefore infinitely many exponential solutions) of a differential-difference equation whereas there are only a finite number of roots of a pure differential equation. This results in a great increase in the complexity of solution processes.

3.6   Laplace Transform Solutions

Laplace Transform methods are extremely useful in obtaining solutions of linear differential-difference equations with constant coefficients. We can illustrate these methods by considering the simple equation

$$(3.6.1) \qquad u'(t) = u(t-1) \qquad .$$

If we multiply this equation by e^{-st} and integrate from 1 to ∞ (proceeding in a purely formal fashion), we get

$$(3.6.2) \qquad \int_1^\infty u'(t)e^{-st}dt = \int_1^\infty u(t-1)e^{-st}dt \qquad .$$

Using a change of variable, we find that

$$\int_1^\infty u(t-1)e^{-st}dt = e^{-s}\int_0^\infty u(t)e^{-st}dt$$

$$= e^{-s}\left[\int_1^\infty u(t)e^{-st}dt + \int_0^1 u(t)e^{-st}dt\right] \qquad .$$

Integrating by parts yields, assuming $u(t)e^{-st} \to 0$ as $t \to \infty$,

$$\int_1^\infty u'(t)e^{-st}dt = -u(1)e^{-s} + s\int_1^\infty u(t)e^{-st}dt \qquad .$$

Thus (3.6.2) takes the form,

$$(s-e^{-s})\int_1^\infty u(t)e^{-st}dt = u(1)e^{-s} + e^{-s}\int_0^1 u(t)e^{-st}dt \qquad .$$

Assuming $(s-e^{-s}) \neq 0$, we obtain

$$(3.6.3) \qquad \int_1^\infty u(t)e^{-st}dt = \frac{u(1)e^{-s} + e^{-s}\int_0^1 u(t)e^{-st}dt}{s-e^{-s}} \qquad ,$$

an equation which expresses the transform of u in terms of the value of u over time interval $0 \le t \le 1$. Assuming that the inversion formula discussed in Chapter 1 can be applied, we get

$$(3.6.4) \qquad u(t) = \frac{1}{2\pi i} \int_{(c)} \frac{\left\{ u(1)e^{-s} + e^{-s} \int_0^1 u(p)e^{-sp}\, dp \right\} e^{st}\, ds}{s - e^{-s}}$$

$$t > 0 \quad ,$$

where (c) is a contour in the complex plane.

Thus, provided the various steps above can be justified rigorously, we see that the solution of $(3.6.1)$ can be expressed in terms of the initial values of $u(t)$ over the interval $(0,1)$ by means of a contour integral.

We can now extend the results to a general first-order linear differential-difference equation,

$$(3.6.5) \qquad L(u) = a_0 u'(t) + b_0 u(t) + b_1 u(t-c) = 0, \qquad c > 0 \quad .$$

The characteristic equation of $(3.6.5)$ is,

$$(3.6.6) \qquad h(s) = a_0 s + b_0 + b_1 e^{-cs} = 0 \quad .$$

It is characteristic of Eq. $(3.6.6)$ that all the roots lie in the complex s-plane to the left of some vertical line. That is, there is a real constant c_0 such that all roots s satisfy the condition $\mathrm{Re}(s) > c_0$. With the location of the characteristic roots of $(3.6.6)$ bounded by c_0, we may characterize the solution $u(t)$ of the inhomogeneous linear differential-difference equation,

$$(3.6.7) \qquad L(u) = a_0 u'(t) + b_0 u(t) + b_1 u(t-c) = f(t),$$

$$t \geq c, \quad a_0 \neq 0$$

satisfying the initial condition $u(t) = g(t)$, $0 \leq t \leq c$. We shall assume that g is $C^0 [0,c]$, that f is $C^0 [0,\infty)$, and that

$$|f(t)| \leq c_1 e^{c_2 t}, \qquad t \geq 0, \qquad c_1 > 0, \qquad c_2 > 0 \quad ,$$

then for any sufficiently large constant c, we shall show that

$$(3.6.8) \qquad u(t) = \frac{1}{2\pi i} \int_{(c)} e^{st} h^{-1}(s)(p_0(s) + q(s))ds \quad , \quad t > c \quad ,$$

where

$$(3.6.9) \qquad p_0(s) = a_0 g(c)e^{-cs} - b_1 e^{-cs} \int_0^c g(t_1)e^{-st_1}dt_1 \quad ,$$

$$(3.6.10) \qquad q(s) = \int_c^\infty f(t_1)e^{-st_1} dt_1 \quad .$$

Also, if g is restricted to be $C^1 [0,c]$, then

$$(3.6.11) \qquad u(t) = \frac{1}{2\pi i} \int_{(c)} e^{st} h^{-1}(s)(p(s) + q(s))ds \quad , \quad t \geq 0 \quad ,$$

where

$$p(s) = a_0 g(c)e^{-cs} + (a_0 s + b_0) \int_0^c g(t_1)e^{-st_1}dt_1 \quad .$$

In the forms (3.6.8) and (3.6.11), the solution of the homogeneous differential-difference Eqs. (3.6.6) is written explicitly as a function of the characteristic equation $h(s)$, the initial conditions, $g(t)$, $0 \leq t \leq c$, and the forcing function $f(t)$. To show that this useful result is valid, we first note that we have placed enough restrictions on g and f to insure the existence of two positive constants c_3 and c_4 such that,

$$(3.6.12) \qquad |u(t)| \leq c_3 e^{c_4 t}, \qquad t > 0 \quad .$$

This result is shown in the next sections.

Hence the integrals

$$\int_0^\infty u(t)e^{-st}dt, \qquad \int_c^\infty u(t-c)e^{-st}dt \quad , \qquad \int_0^\infty f(t)e^{-st}dt$$

converge for any complex number s for which $\text{Re}(s) < c_4$. By integration by parts, we get,

$$(3.6.13) \quad \int_c^t u'(t_1)e^{-st_1}dt_1 = u(t)e^{-st} - g(c)e^{-cs} + s\int_c^t u(t_1)e^{-st_1}dt_1$$

Since $u(t)e^{-st}$ approaches zero as $t \to \infty$ if $\text{Re}(s) > c_4$, by virtue of (3.6.12), the right-hand member of (3.6.13) converges as $t \to \infty$. Hence the left-hand member also converges and

$$(3.6.14) \quad \int_c^\infty u'(t)e^{-st}dt = -g(c)e^{-cs} + s\int_c^\infty u(t)e^{-st}dt \quad .$$

Moreover it is clear that,

$$(3.6.15) \quad \int_c^\infty u(t-c)e^{-st}dt = e^{-cs}\left[\int_c^\infty u(t)e^{-st}dt + \int_0^c g(t)e^{-st}dt\right] \quad .$$

Substituting into Eq. (3.6.7) after integrating both sides from c to infinity, we obtain,

$$(3.6.16) \quad h(s)\int_c^\infty u(t)e^{-st}dt = p_0(s) + q(s), \qquad \text{Re}(s) > c_4 \quad .$$

Since we have shown that $h(s)$ is not zero for $\text{Re}(s) > c$, if c is sufficiently large, we can write,

$$(3.6.17) \quad \int_c^\infty u(t)e^{-st}dt = h^{-1}(s)(p_0(s) + q(s)), \qquad \text{Re}(s) > c \quad .$$

Since $u(t)$ is of class C^1 on $[c, \infty)$, and is continuous and of bounded variation on any finite interval, we can employ the inversion formula for the Laplace Transform to obtain the equation in (3.6.8).

To obtain the equation in (3.6.11), we alter the procedure slightly. Instead of using the equations in (3.6.14) and (3.6.15), we use

$$\int_c^\infty u'(t)e^{-st}dt = -g(c)e^{-cs} + s\left[\int_0^\infty u(t)e^{-st}dt\right.$$

$$\left. - \int_0^c g(t)e^{-st}dt\right]$$

$$\int_c^\infty u(t-c)e^{-st}dt = e^{-st}\int_0^\infty u(t)e^{-st}dt \quad .$$

When we substitute into the transformed Eq. (3.6.7) as before and solve for $\int_c^\infty u(t)e^{-st}dt$, we obtain,

$$(3.6.18) \qquad \int_0^\infty u(t)e^{-st}dt = -h^{-1}(s)(p(s) + q(s)), \qquad \text{Re}(s) > c \quad .$$

Since $g(t)$ is assumed to be $C^1[0,c]$, $u(t)$ is C^1 on $[0,c]$ and $[c,\infty)$, thus the inversion formula yields the results in (3.6.11).

It is instructive at this point to make a few remarks about the distinction between equations of the retarded type and equations of advanced type. The equation in (3.6.7) is of the former type, while the equation

$$(3.6.19) \qquad a_1 u'(t-c) + b_0 u(t) + b_1 u(t-c) \text{ f } f(t)$$

is of the latter type. If we apply the above methods to the latter equation, we are led to a characteristic equation of the form,

$$(3.6.20) \qquad h(s) = a_1 s e^{-cs} + b_0 + b_1 e^{-cs} \quad .$$

It can be shown that in this case the characteristic equation has zeros of arbitrarily large real part. For this reason, $h^{-1}(s)$ has singularities for values of s with arbitrarily large real part and the previous procedure fails. As a matter of fact, since each characteristic root gives rise to a solution of (3.6.19), the solutions are not exponentially bounded as in (3.6.11), so that the infinite integrals written above, diverge. It is clear that the procedure must be extensively modified before it can be applied to equations of advanced type.

3.7   Order Of Growth Of Solutions

Since we are primarily interested in the Laplace Transform technique in solving differential-difference equations, it is convenient to have some *a priori* estimate of the magnitude of solutions to these equations. To this end, we shall consider the following results.

If $w(t)$ is continuous, positive and monotone nondecreasing and if

(a) $u(t)$ and $v(t) \geq 0$ are continuous,

(b) $u(t) \leq w(t) + \int_a^t u(s)v(s)ds$, $a \leq t \leq b$,

then

$$u(t) \leq w(t)e^{\int_a^t v(s)ds} .$$

To show this result, we observe,

(3.7.1) $u(t)/w(t) \leq 1 + \int_a^t (u(s)v(s)/w(t))ds$

$\leq 1 + \int_a^t (u(s)v(s)/w(s))ds$.

Now we have the following result at our disposal. If

$$c_1 \geq 0, \quad u(t) \geq 0, \quad v(t) \geq 0 \quad ,$$

the inequality

$$(3.7.2) \qquad u(t) \leq c_1 + \int_0^t u(s)v(s)ds$$

implies,

$$(3.7.3) \qquad u(t) \leq c_1 e^{\int_0^t v(s)ds} \quad .$$

To show this, we observe,

$$\frac{u(t)v(t)}{c_1 + \int_0^t u(s)v(s)ds} \leq v(t) \quad .$$

Integrating both sides over $(0,t)$,

$$(3.7.4) \qquad \int_0^t \frac{u(s)v(s)ds}{c_1 + \int_0^t u(z)v(z)dz} \leq \int_0^t v(s)ds \quad .$$

Set $x(t) = c_1 + \int_0^t u(s)v(s)ds$

then $dx = u(t)v(t)dt$, and (3.7.4) can be written as

$$(3.7.5) \qquad \int_0^t dx/x = \log x \Big|_0^t \leq \int_0^t v(s)ds \quad .$$

Consequently,

$$(3.7.6) \qquad \log\left(c_1 + \int_0^t u(s)v(s)ds\right) - \log c_1 \leq \int_0^t v(s)ds \quad ,$$

or

$$(3.7.7) \qquad c_1 + \int_0^t u(s)v(s)ds \leq c_1 e^{\int_0^t v(s)ds} \quad ,$$

and finally using the hypotheses, we get

$$(3.7.8) \qquad u(t) \leq c_1 e^{\int_0^t v(s)ds} \quad .$$

Now if we set $\bar{u}(t) = u(t)/w(t)$, and let $c_1 = 1$, we have the result,

$$(3.7.9) \qquad u(t)/w(t) \leq e^{\int_0^t v(s)ds}$$

and finally we can say,

$$(3.7.10) \qquad u(t) \leq w(t)e^{\int_0^t v(s)ds} \quad .$$

Now we are in a position to estimate a bound on the solution $u(t)$ of the differential-difference equation,

$$(3.7.11) \qquad L(u) = a_0 u'(t) + b_0 u(t) + b_1 u(t-c) = f(t)$$

where u is of class C^1 on $[0,\infty)$. Let $f(t)$ be of class C^0 on $[0,\infty)$ and assume

$$|f(t)| \leq c_1 e^{c_2 t}, \qquad t \geq 0$$

where c_1 and c_2 are positive constants.
Let

$$m = \max |u(t)|, \qquad 0 \leq t \leq c \quad .$$

Then we wish to show that there exists two constants c_3 and c_4, depending on c_2, a_0, b_0, b_1, such that,

$$(3.7.12) \qquad |u(t)| \leq c_3(c_1 + m)e^{c_4 t}$$

We begin by integrating $L(u)$, therefore we have the relation

(3.7.13) $a_0 u(t) = a_0 u(c) + \int_c^t f(s)ds - b_0 \int_c^t u(s)ds$

$$- b_1 \int_c^t u(s-c)ds \quad .$$

By taking the norm of both sides, we get the inequality,

(3.7.14) $|a_0 u(t)| \leq |a_0|m + c_1 \int_c^t e^{c_2 s} ds + |b_0| \int_0^t |u(s)|ds$

$$+ |b_1| \int_0^{t-c} |u(s)|ds \quad t \geq c \quad ,$$

or

(3.7.15) $|u(t)| \leq m + \dfrac{c_1}{c_2|a_0|} e^{c_2 t} + \dfrac{|b_0| + |b_1|}{|a_0|} \int_0^t |u(s)|ds \quad .$

Set

$$c_3 = \max(1, \; 1/c_2|a_0|)$$

$$c_5 = \frac{|b_0| + |b_1|}{|a_0|}$$

then

(3.7.16) $|u(t)| \leq c_3(c_1 + m)e^{c_2 t} + c_5 \int_0^t |u(s)|ds \quad .$

Since $|u(t)| \leq m \leq c_3 m e^{c_2 t}$, for $0 \leq t \leq c$, the inequality (3.7.16) holds for all $t \geq 0$.
Let

$$w(t) = c_3(c_1 + m)e^{c_2 t}, \qquad v(t) = e^{c_5 t} \quad ,$$

then

(3.7.17) $|u(t)| \leq c_3(c_1+m)e^{c_2 t}\int_0^t |e^{c_2 t}|ds$,

and finally

(3.7.18) $|u(t)| \leq c_3(c_1+m)e^{(c_2+c_3)t}$

thereby giving an asymptotic bound to the function $u(t)$.

3.8   The Characteristic Roots

We have shown that under suitable conditions, the solution $u(t)$ of

(3.8.1) $a_0 u'(t) + b_0 u(t) + b_1 u(t-c) = 0$, $t > c$, $a_0 \neq 0$

(3.8.2) $u(t) = f(t)$, $0 \leq t \leq c$,

is given by

(3.8.3) $u(t) = \dfrac{1}{2\pi i}\displaystyle\int_{(c)} e^{st}h^{-1}(s)p(s)ds$, $t > 0$,

where $h(s)$ is the characteristic function associated with (3.8.1) and $p(s)$ reflect the initial conditions. For many purposes, it is of great value to obtain a representation of $u(t)$ in the form of an infinite series. It is easy to see how such an expression can arise, for if it is possible to deform the line of integration in (3.8.3) into a contour surrounding all the zeros of $h(s)$, the residue theorem would at once yield the relation,

u(t) = sum of the residue of $(e^{st}h^{-1}(s)p(s))$.

This suggests that we may be able to establish expansions of the form

(3.8.4) $u(t) = \sum p_r(t)e^{s_r t}$,

where the sum is over all the characteristic roots s_r and where

$pr(t)$ is a polynomial in t if s_r is a multiple root.

In order to determine some of the characteristics of the solution $u(t)$ of (3.8.1) it is clearly necessary to have a good idea of the location of the zeros of $h(s)$, where

$$(3.8.5) \qquad h(s) = a_0 s + b_1 e^{-cs}$$

approximates the characteristic function of (3.8.1) as $|s| \to \infty$. The zeros of (3.8.5) are values of s for which

$$(3.8.6) \qquad a_0 s + b_1 e^{-cs} = 0, \qquad c > 0 \quad .$$

Therefore it must be true that for all zeros,

$$(3.8.7) \qquad |s e^{cs}| = |b_1/a_0| \quad ,$$

or by taking logs of both sides,

$$(3.8.8) \qquad Re(s) + c^{-1} \log|s| = c^{-1} \log|b_1/a_0| \, , \qquad b_1 \neq 0 \quad .$$

It is instructive to let $s = x + iy$, thus (3.8.8) becomes,

$$(3.8.9) \qquad x + c^{-1} \log((x^2 + y^2)^{\frac{1}{2}}) = c^{-1} \log|b_1/a_0| \quad .$$

Equation (3.8.9) defines a continuous curve in the complex s-plane along which the zeros of (3.8.6) lie. Furthermore, it can be shown that as $s \to \infty$, the roots of $h(s)$ lie asymptotically along this curve.

By examining (3.8.9), several facts become clear. First, we observe that the curve is symmetric with respect to the x-axis. Next we can rewrite (3.8.9) in a more illuminating form,

$$(3.8.10) \qquad e^{cx} = \frac{|b_1/a_0|}{\sqrt{x^2 + y^2}} \, , \qquad c > 0 \quad .$$

If we assume $x > 0$, then for $x > x_0$, depending on b_1/a_0 and c, there is no real value of y to satisfy (3.8.10). On the other hand, if $x < 0$, we can always find a y satisfying (3.8.10).

Therefore the curve defined in (3.8.10) lies entirely within a left half of the complex s-plane. Furthermore, as $|s| = (x^2 + y^2)^{\frac{1}{2}}$ becomes large, the curve becomes more nearly parallel to the imaginary axis and Re(s).

Moreover the asymptotic location of the zeros of h(s) can be very precisely described. For example, it can be shown that there is a constant, c > 0 such that all zeros of sufficiently large modulus lie within a region defined by the inequalities,

$$(3.8.11) \quad -c \leq Re(s) + c^{-1} \log|s| \leq c \quad .$$

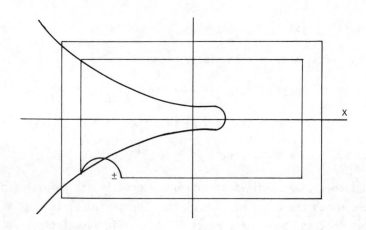

Fig. 3.1 Asymptotic location of the zeros of h(s).

Thus we see that in principle it is possible to represent the solution (3.8.6) of an equation of the retarded type (3.8.1) as a sum of exponentials. Carrying out such a computation is, however, quite difficult to do in practice. Indeed, all zeros for which s is sufficiently large have the form

$$(3.8.12) \quad s = c^{-1} \left(\log|b_1/a_0| - \log\left|\frac{c_1 + 2\pi k}{c}\right| + c_1/\pi k + O(1/k) \right)$$

$$\pm \frac{i}{c} (c_1 + 2\pi k + C(\log k/k)) \quad ,$$

where k represents any large positive integer and c_1 is $\pm \pi/2$ according as $a_0^{-1}b_1$ is positive or negative. This means that the zeros are spaced along the curve with an asymptotic distance of $2\pi/c$ apart and moreover there exists a sequence of closed contours C_ℓ ($\ell = 1,2,...$) in the complex plane, and a positive integer ℓ_0 with the following properties:

(a) C_1 contains the origin as an interior point.

(b) C_ℓ is contained in the interior of $C_{\ell+1}$ ($\ell = 1,2,...$).

(c) The contours C_ℓ have a least distance $d > 0$ from the set of all zeros of $h(s)$. That is, when s lies on a contour and s_n is a zero,

$$\inf_{s,s_n} |s - s_n| = d > 0 \ .$$

(d) The contour C_1 lies along the circle $|s| = (\ell)\pi/c$ outside of the strip given by (3.8.7). Inside this region it lies between the circle $|s| = (\ell-1)\pi/c$ and the circle $(\ell+1)\pi/c$.

(e) The portion of C_ℓ within the region is of a length which is bounded as $\ell \to +\infty$.

(f) For $\ell > \ell_0$ there is exactly one zero of $h(s)$ between C_ℓ and $C_{\ell+1}$.

It is perhaps appropriate to make a few remarks here about the essential difference between equations of retarded, neutral and advanced types. The nature of the distribution of the zeros of the characteristic function has been described above for the retarded equation (3.4.1). For the neutral equation

$$(3.8.13) \qquad a_0 u'(t) + a_1 u'(t-c) + b_0 u(t) + b_1 u(t-c) = 0 \ ,$$

$$a_0, \ a_1 \neq 0 \quad ,$$

the characteristic function is

$$(3.8.14) \qquad h(s) = a_0 s + a_1 s e^{-cs} + b_0 + b_1 e^{-cs} \quad ,$$

and the zeros are asymptotically distributed along a vertical line. Since the zeros thus fail to have the property that $\mathrm{Re}(s) \to -\infty$ as we follow along the curve, the nature of the solution is considerably altered. For the advanced type, whose equation is,

$$(3.8.15) \qquad a_1 u'(t-c) + b_0 u(t) + b_1 u(t-c) = 0, \qquad a_1 \neq 0 \quad ,$$

the characteristic function

$$(3.8.16) \qquad h(s) = a_1 s e^{-cs} + b_0 + b_1 e^{-cs}$$

has zeros asymptotically distributed along a curve on which $\mathrm{Re}(s) \to +\infty$.

3.9   Asymptotic Behavior of the Solution

One of the important problems in the study of differential equations or of differential-difference equations and their applications is that of describing the nature of the solutions for large positive values of the independent variable. It is clear from the results of the last section that the nature of the solution for large t is closely related to the distribution of the characteristic roots in the complex plane. It is the purpose of this section to explore this connection in greater detail. Particular attention will be given to the problem of finding conditions under which a solution approaches zero as $t \to \infty$, or is "very small" for all t, or is bounded as $t \to \infty$.

The general method used here is quite simple. It has been shown that for any positive integer, R,

$$(3.9.1) \qquad u(t) = \sum_{r=1}^{R} e^{s_r t} p_r(t) + \text{error}$$

where the error is the sum of terms of exponential order which decreases to zero. The larger we take R, the smaller the error becomes. It, therefore, seems plausible that for large t, $u(t)$ is closely approximated by terms in the sum of (3.9.1). In this

77

section we shall give a rigorous development of this idea.

From (3.6.11) we have,

$$(3.9.2) \qquad u(t) = \frac{1}{2\pi i} \int_{(c)} e^{st} h^{-1}(s) p(s) ds, \qquad t > 0 \quad ,$$

or (3.6.8),

$$(3.9.3) \qquad u(t) = \frac{1}{2\pi i} \int_{(c)} e^{st} h^{-1}(s) p_0(s) ds, \qquad t > c$$

for any c which exceeds the largest of the real parts of the characteristic roots. Since the last section was devoted to locating the zeros of $h(s)$ (or the poles of $h(s)^{-1}$) in the complex plane, we can find the asymptotic behavior of the solution of the differential-difference equations of the retarded type. The zeros of $h(s)$ has been shown to be distributed in the complex s-plane as shown in Fig. 3.2.

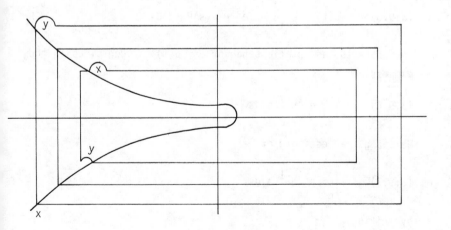

Fig. 3.2 Location of the zeros of the
characteristic function $h(s)$

By the construction of the contours given in the last section, we know that exactly one zero of $h(s)$ lies between contours c_ℓ and $c_{\ell+1}$.

Let us consider carefully shifting the contour c defined in (3.9.3) to the left to a line $\text{Re}(s) = c'$, $c' < c$, on which no characteristic roots lie. To do this we consider the contour c_ℓ, for a given ℓ, and define a closed contour K_ℓ, whose vertical sides comprise portions of c and c' and whose horizontal segments are those of c_ℓ. As $\ell \to \infty$ the contour K_ℓ defines an infinite strip in the complex s-plane and the integrands over the horizontal parts of K_ℓ approach zero.

Therefore (3.9.3) can be rewritten as

$$(3.9.4) \qquad u(t) = \lim_{\ell \to \infty} \left[\frac{1}{2\pi i} \int_{(c)} e^{st} h^{-1}(s) P_0(s) ds + \sum e^{s_r t} P_r(t) \right],$$

$$t > c$$

where s_r are the poles of $h^{-1}(s)$ lying within the strip defined by K_ℓ and $P_r(t)$ is a polynomial dictated by the order of the poles. Since there are, at most, a finite number of roots of $h(s)$ in the strip, we have

$$(3.9.5) \qquad u(t) = \frac{1}{2\pi i} \int_{c'} e^{st} h^{-1}(s) P_0(s) ds + \sum_{\text{Re}(s_r) > c'} e^{s_r t} P_r(t)$$

therefore the asymptotic behavior of $u(t)$ is contained in the expression,

$$(3.9.6) \qquad \int_{c'} e^{st} h^{-1}(s) ds \quad .$$

We can get an estimate of

$$(3.9.7) \qquad \int_{c'} e^{st} h^{-1}(s) ds$$

by writing

$$(3.9.8) \qquad \int_{c'} e^{st} h^{-1}(s) ds = \int_{c'} \frac{e^{st}}{a_0 s + b_0} ds$$

$$- b_1 \int_{c'} \frac{e^{(t-c)s}}{(a_0 s + b_0) h(s)} ds \quad .$$

The first integral in the right-hand member of (3.9.8) has well-known convergence properties. The second is uniformly convergent for t in any finite interval. Since $|h^{-1}(s)| = 0(|s|^{-1})$ and for $t \geq 0$ if $c' < 0$, therefore (3.9.8) is $0(e^{c't})$ as $t \to +\infty$.

From the known form of $P_0(s)$, we can show that

$$(3.9.9) \qquad \frac{1}{2\pi i} \int_{c'} e^{st} h^{-1}(s) P_0(s) ds = 0(e^{c't}) \quad \text{as} \quad t \to \infty \quad .$$

We can summarize this result by saying if $g(t)$ is of class c^0 on $[0,c]$ and $u(t)$ is a continuous solution of the equation,

$$(3.9.10) \qquad a_0 u'(t) + b_0 u(t) + b_1 u(t-c) = 0, \qquad t > c, \qquad a_0 \neq 0 \quad ,$$

satisfying the initial conditions, $u(t) = g(t)$, $0 \leq t \leq c$, and we let $e^{s_r t} p_r(t)$ denote the residues of $e^{st} h^{-1}(s) P_0(s)$ at a zero s_r of $h(s)$.

In addition, if we let

$$m_g = \max |g(t)|, \qquad 0 \leq t \leq c \quad ,$$

and c be any number such that no zeros of $h(s)$ lie on the line $\text{Re}(s) = c$, then there exists a positive number c_1, independent of t and $g(t)$, such that,

$$(3.9.11) \qquad |u(t) - \sum_{\text{Re}(s_r) > c} e^{s_r t} p_r(t)| \leq c_1 m_g e^{ct} \quad ,$$

where the sum is taken over all characteristic roots s_r to the right of the line $\text{Re}(s) = c$.

From this result we can conclude the conditions under which the solutions of (3.9.10) approach zero as $t \to \infty$ and the conditions under which they are all bounded.

A necessary and sufficient condition in order that all

continuous solutions of Eq. (3.9.10) approach zero as $t \to \infty$ is that all characteristic roots of $h(s)$ have negative real parts.

A necessary and sufficient condition in order that all continuous solutions of Eq. (3.9.11) be bounded as $t \to \infty$, is that

(a) all characteristic roots of $h(s)$ have nonpositive real parts,

(b) if s_r is a root with zero real part, the residue of $e^{st}h^{-1}(s)$ at s_r be bounded as $t \to \infty$.

A necessary and sufficient condition for condition (b) to be satisfied, is that each root with zero real part be simple.

3.10   The Shift Theorem

We have found that we can obtain a solution of a differential-difference equation in the form of a contour integral or a series of exponentials. We have also considered the behavior of the solution for large values of t. However, if we are interested in small or intermediate values of t, the fundamental method of continuation is still of prime importance. In this section, we shall show how Laplace Transform methods can sometimes be useful in providing this continuation. To illustrate the procedure, we shall consider the equation,

(3.10.1) $u'(t) = u(t - 1), \qquad t > 1$

with the initial condition,

(3.10.2) $u(t) = 1, \qquad 0 < t < 1$.

From our previous results, we know that the characteristic functio associated with (3.10.1) is,

(3.10.3) $h(s) = s - e^{-s}$.

The solution of (3.10.1), when written in terms of a contour integral is given as,

(3.10.4) $u(t) = \dfrac{1}{2\pi i} \displaystyle\int_C e^{st} h^{-1}(s) p(s) ds$,

where

$$p(s) = e^{-s} + s \int_s^1 e^{-st} dt \quad .$$

Formally we can now compute,

$$h(s) \int_0^\infty u(t) e^{-st} dt = h(s) \int_0^\infty e^{-st} \int_C e^{st} h^{-1}(s) p(s) ds \, dt$$

$$= \int_0^\infty \int_C p(s) ds \, dt$$

$$= p(s)$$

$$= e^{-st} + \int_0^1 e^{-st} dt$$

$$= e^{-st} - (e^{-st} - 1)$$

$$= 1 \quad .$$

Using this result we can eliminate $u(t)$ from the expression,

(3.10.5) $\displaystyle\int_0^\infty u(t) e^{-st} dt = 1/h(s)$

$$= 1/(s - e^{-s})$$

$$= 1/(s(1 - e^{-s}/s))$$

but we know $1/(1 - e^{-s}/s)$ has the infinite expansion,

(3.10.6) $1/(1 - e^{-s}/s) = \displaystyle\sum_{n=0}^\infty (e^{-s}/s)^n \quad .$

Combining these results, we can say that, provided $Re(s)$ is sufficiently large,

(3.10.7) $\displaystyle\int_0^\infty u(t) e^{-st} dt = \sum_{n=0}^\infty s^{-(n+1)} e^{-ns} \quad .$

We now use the well-known fact that

$$(3.10.8) \qquad \int_0^\infty t^n e^{-st} dt = n!/s^{n+1} \quad ,$$

together with the exponential shift theorem, which states that if $H(t)$ is the unit step function, then,

$$(3.10.9) \qquad \int_0^\infty u(t-c)H(t-c)e^{-st} dt = e^{-cs} \int_0^\infty u(t)e^{-st} dt \quad , \qquad c > 0$$

From relations (3.10.8) and (3.10.9), it follows that,

$$(3.10.10) \qquad \int_0^\infty (t-c)^n H(t-c)e^{-st} dt = e^{-cs} n!/s^{n+1} \quad .$$

Now set $c = n$, and rewrite (3.10.10)

$$(3.10.11) \qquad \int_0^\infty \sum_{n=0}^\infty (t-n)^n H(t-n)/n! e^{-st} dt = \sum_{n=0}^\infty e^{-ns}/s^{n+1} \quad .$$

By the uniqueness of the Laplace Transform,

$$(3.10.12) \qquad u(t) = \sum_{n=0}^\infty (t-n)^n H(t-n)/n! \quad .$$

The above method is applicable whenever the function $u(t)$ can be expressed as a combination of functions whose inverse transforms are known. In the case where this is not true, it may still be possible to use similar expansion methods to advantage.

3.11   Equations of the Neutral and Advanced Types

In this section, we will continue some of the results for linear differential-difference equations of neutral or advanced type. These results are analogous to the results of the last section. We shall first consider the properties of a first-order equation of the neutral type, i.e.,

$$(3.11.1) \qquad a_0 u'(t) + a_1 u'(t-c) + b_0 u(t) + b_1 u(t-c) = f(t) \quad,$$

$$a_0, \, a_1 \neq 0 \quad.$$

As usual, the initial condition is of the form,

$$(3.11.2) \qquad u(t) = g(t), \qquad 0 \leq t \leq c \quad.$$

The principal result which is true for (3.11.1) is the following: suppose that f is of class C^0 on $[0,\infty)$ and that g is of class C^1 on $[0,c]$. Then there exists one and only one function $u(t)$ for $t \geq 0$ which is continuous for $t \geq 0$, which satisfies the initial conditions in (3.11.2) and which satisfies the equation in (3.11.1), on each of the intervals $[jc, (j+1)c]$ $j = 1,2,\ldots$ This function is of class C^1 on the intervals $[jc, (j+1)c]$ $j = 0,1,2,\ldots$ It is of class C^1 on $[0,\infty)$ and satisfies the equation in (3.11.1) on $[c,\infty)$, if and only if it has a continuous derivative at $t = c$. This is true if and only if,

$$(3.11.3) \qquad a_0 g'(c-0) + a_1 g'(0) + b_0 g(c) + b_1 g(0) = f(c) \quad.$$

It is interesting to compare this result with the similar result for the equation of the retarded type. An equation of the retarded type seems to smooth out irregularities in the initial values, that is, if the initial function is merely at C^0 then the solution is eventually of class C^n for any n for which f is of class C^{n-1}. On the other hand, an equation of the neutral type has no such smoothing effect. The solution retains the degree of regularity of its initial values.

Let us next consider the first-order equation of the advanced type. To do this we must extend our definition of the classes C^k as follows.

A function f is said to be of class C^∞ on (t_1, t_2) if it possesses continuous derivatives of all orders in the open interval $t_1 < t < t_2$. It is said to be of class C^∞ on $[t_1, t_2)$ if it is of class C^∞ on (t_1, t_2) and if for $k = 0,1,2,\ldots$ it has a right-hand k-th derivative at t_1 and the function $f^{(k)}(t)$ defined over $t_1 \leq t \leq t_2$ by these values is continuous from the right at t_1. It is of class C^∞ on $(t_1, t_2]$ if these statements are valid when "right", is replaced by "left" and "t_1" by "t_2". It is of class C^∞ on $[t_1, t_2]$ if it is of class C^∞ on $[t_1, t_2)$ and $(t_1, t_2]$.

With the aid of the above definitions, the following result can be proven. Suppose that f is of class C^∞ on $[0,\infty)$ and that g is of class C^∞ on $[0,c]$. Then there exists one and only one function $u(t)$ which satisfies the initial condition on (3.11.2) and which satisfies the equation

$$(3.11.4) \qquad a_1 u'(t-c) + b_0 u(t) + b_1 u(t-c) = f(t), \qquad a_1 \neq 0,$$

$$b_0 \neq 0$$

on each of the intervals $[jc, (j+1)c]$, $j = 1, 2, \ldots$ This function is of class C^∞ on $[0,\infty)$, and satisfies the equation (3.11.4) on $[c,\infty)$, if it is continuous and has continuous derivatives of all orders at $t = c$. This is true if and only if all the relations,

$$(3.11.5) \qquad a_1 g^{(k+1)}(0+) + b_0 g^{(k)}(c-0) + b_1 g^{(k)}(0+) = f^{(k)}(c) \quad ,$$

$$k = 0, 1, 2, \ldots$$

are satisfied.

With the existence of solutions assured, we can say something about the type of solutions which might be expected. The validity of exponential solutions for both the neutral and advanced type of differential-difference equations remains true. The linear operator now has the form,

$$(3.11.6) \qquad L(u) = a_0 u'(t) + a_1 u'(t-c) + b_0 u(t) + b_1 u(t-c) \quad ,$$

and has the characteristic function,

$$(3.11.7) \qquad h(s) = a_0 s + a_1 s e^{-cs} + b_0 + b_1 e^{-cs} \quad .$$

For equations of the neutral type, it is again possible to obtain an exponential bound on the magnitude of solutions, as was done for equations of the retarded type. No such bound can exist for equations of advanced type, since such equations have characteristic roots of arbitrary large real parts. It follows that the Laplace Transform method can be used for equations of the neutral type but not for equations of the advanced type.

3.12   Linear Systems Of Differential-Difference Equations

In this section, we shall show that, with the aid of an appropriate notation, it is fairly easy to extend most of our previous work to much more general equations of the form,

$$(3.12.1) \qquad \sum_{i=0}^{m} \sum_{j=0}^{n} a_{ij} u^{(j)}(t - c_i) = f(t) \qquad ,$$

where m and n are positive integers, where $0 = c_0 < c_1 < \ldots < c_m$, and where $f(t)$ is a given function. Still more general is the linear system of equations

$$(3.12.2) \qquad \sum_{i=0}^{m} \sum_{j=0}^{n} a_{ijk} u_j'(t - c_i) + \sum_{i=0}^{m} \sum_{j=0}^{n} b_{ijk} u_j(t - c_i) = f_k(t)$$

$$k = 1, 2, \ldots, n \ ,$$

involving n unknown functions $u_1(t)$, $u_2(t), \ldots, u_n(t)$, and their first derivatives. Indeed we shall show that Eq. (3.12.1) can be transformed into a system which is of the type shown in (3.12.2). To do this, define new variables v_1, $v_2, \ldots, v_n$, by means of the relations $v_{j+1}(t) = u^{(j)}(t)$ $(j = 0, 1, 2, \ldots, n-1)$.

Then Eq. (3.12.1) is equivalent to the system

$$v_1'(t) - v_2(t) = 0 \ ,$$
$$\vdots$$
$$v_{n-1}'(t) - v_n(t) = 0 \ ,$$

$$\sum_{i=0}^{m} a_{in} v_n'(t - c_i) + \sum_{i=0}^{m} \sum_{j=0}^{n} a_{i,j-i} v_j(t - c_i) = f(t) \quad .$$

We see that this system is a special case of that in (3.12.2).

With the aid of the vector-matrix notation, we can write the general system of equations in (3.12.2) in a simplified form. Indeed, if we let $\overline{y}$ and $\overline{f}$ represent the column vectors,

$$(3.12.3) \quad \overline{y} = \begin{bmatrix} u_1 \\ \cdot \\ \cdot \\ \cdot \\ u_n \end{bmatrix}, \qquad \overline{f} = \begin{bmatrix} f_1 \\ \cdot \\ \cdot \\ \cdot \\ f_n \end{bmatrix}$$

respectively, let A_i be the matrix, the kj-th element of which is a_{jk} and let B_i be the matrix, the kj-th element of which is b_{ijk}. We see at once that the system takes the form,

$$(3.12.4) \quad \sum_{i=0}^{m} A_i \overline{y}'(t - c_i) + \sum_{i=0}^{m} B_i \overline{y}(t - c_i) = \overline{f} \quad .$$

As in the case of the scalar equation, we would expect the vector system of differential-difference equations to exhibit properties similar to those of the retarded, neutral, and advanced type. However, it is not immediately clear how to classify such systems in an appropriate fashion. One way to shed light on this matter is to examine the distribution of the roots of the characteristic function. This equation can be shown to be,

$$(3.12.5) \quad \det\left(\sum_{i=0}^{m} (A_i s + B_i) e^{-c_i s} \right) = 0 \quad .$$

The roots of large magnitude of this equation are grouped in a finite number of chains. Each chain of roots lies asymptotically along a certain curve of one of the kinds encountered in the scalar case. A chain of roots may be said to be a retarded chain if $Re(s) \to -\infty$ as $|s| \to \infty$, a neutral chain if $Re(s)$ is bounded as $|s| \to \infty$ and an advanced chain if $Re(s) \to \infty$ as $|s| \to \infty$. The

existence and uniqueness of the solution to such systems of
differential-difference equations are embodied in the following
result.

Consider the system of differential-difference equations
with constant coefficients,

$$(3.12.6) \qquad \sum_{i=0}^{m} \left[A_i \bar{y}'(t - c_i) + B_i \bar{y}(t - c_i) \right] = \bar{f}(t)$$

$$0 = c_0 < c_1 < c_2 < \ldots < c_m , \qquad \det A_0 \neq 0 ,$$

and the initial conditions,

$$(3.12.7) \qquad \bar{y}(t) = \bar{g}(t), \qquad 0 \le t \le c_m .$$

Let S denote the set of points of the form

$$(3.12.8) \qquad t_m = \sum_{i=0}^{m} j_i c_i ,$$

where j_i are integers. Let S_1 denote the intersection of S
with $[c_m, \infty)$ and let S_2 denote the intersection of S with
(c_m, ∞). Suppose that the vector g is of class C^1 on $[0, c_m]$
and that the vector f is of class C^0 on $[0, \infty)$ except for
possible finite jump discontinuities at points of the set S_1.
Then there exists one and only one vector function $\bar{y}(t)$ for $t \ge 0$
which is continuous for $t \ge 0$, which satisfies the initial condition in (3.12.7) and which satisfies Eq. (3.12.6) for $t > c_m$,
$t \neq S_2$. Furthermore, $y(t)$ is of class C^1 on $[0, \infty)$ and
satisfies (3.12.6) for all $t > c_m$, if and only if it has a
continuous derivative at $t = c_m$. This is true if and only if,

$$(3.12.9) \qquad \sum_{i=0}^{m} A_i g'(c_m - c_i) + B_i g(c_m - c_i) = f(c_m) .$$

Suppose that f is of class C^1 on $[0, \infty)$ and that g
is of class C^2 on $[0, c_m]$. Then $y(t)$ is of class C^2 for

$t > 0$, $t \neq S_1$. A sufficient condition that it be of class C^2 on $|0, \infty)$ is that Eq. (3.12.9) and the equation

$$(3.12.10) \quad \sum_{i=0}^{m} A_i g''(c_m - c_i) + B_i g'(c_m - c_i) = f'(c_m)$$

is satisfied. In proving that this result is true, we may proceed in the following manner. Since A_0 is nonsingular, we can multiply (3.12.6) by A_0^{-1}, or simply set $A_0 = I_1$. Now let,

$$(3.12.11) \quad v(t) = f(t) - \sum_{i=1}^{m} (A_i y'(t - c_i) + B_i y(t - c_i)) \quad .$$

Hence, we now have $y'(t) + B_0 y(t) = v(t)$ and the system can be rewritten as,

$$(3.12.12) \quad d/dt(e^{B_0 t} y(t)) = e^{B_0 t} v(t) \quad .$$

Since $f(t)$ is of class C^0 on $[0, \infty)$ except on S_1 and $g(t)$ is of class C^1 on $[0, c_m]$, $v(t)$ is of class C^0 on $(c_m, c_m + c_1)$ Then by integration of (3.12.12), we have a unique function $y(t)$ satisfying (3.12.6) on $[c_m, c_m + c_1]$ and $y(t) = g(t)$ on $0 \leq t \leq c_m$ This function is of class C^1 on $[0, c_m + c_1]$ except for a possible jump in $y'(t)$ at $t = c_m$. It follows that $v(t)$ is of class C^0 on $[c_m, c_m + 2c_1]$ except for possible jumps at points of the form $t = c_m + c_j$, $j = 1, 2, \ldots, m$, $t < c_m + 2c_1$.

Repeating the use of the relation in (3.12.11), we see that $v(t)$ is of class C^0 in $[c_m, c_m + 3c_1]$ except for possible jumps of the form $t = c_m + c_j + c_k$, $(j = 1, 2, \ldots, m, k = 0, 1, \ldots, m) t < c_m + 3c_1$, or in other words, at the points of the form $t_m = \sum_{i=1}^{m} j_i c_i$, $t < c_m + 3c_1$, where j_i are nonnegative integers, $j_m \geq 1$ and $2 \leq \sum_{i=1}^{m} j_i \leq 3$. By integration of (3.12.12) we can continue $y(t)$

over $[0, c_m + 3c_1]$. This process can be continued indefinitely establishing the existence of a unique function $y(t)$ of class C^0 on $[0,\infty]$ satisfying (3.12.6) and the initial conditions (3.12.7) for $t > c_m$, $t \notin S_2$. This function is of class C^1 for $t > 0$ with possible jumps in the derivatives at $t \in S_1$.

It should be clear from our discussion so far that if f is of class C^0 on $[0,\infty)$, then y is of class C^1 on $[0,\infty)$ and satisfies (3.12.6) for all $t > c_m$, if and only if it has a continuous derivative at $t = c_m$. This is true if and only if $y'(c_m + 0) = g'(c_m - 0)$, i.e., if and only if (3.12.9) holds.

Now suppose f is of class C^1 on $[0,\infty)$ and g is of class C^2 on $[0,\infty)$. From (3.12.6) we have

$$(3.12.13) \quad y'(t) = f(t) - B_0 y(t) - \sum_{i=1}^{m} (A_i y'(t - c_i) + B_i y(t - c_i))$$

for $t > c_m$, $t \notin S_1$. Since the right-hand member of the equation has continuous derivatives on $[c_m, c_m + c_1]$, $y(t)$ is of class C^2 on $[c_m, c_m + c_1]$. By repeating this argument for one interval after the other we find that $y(t)$ is C^2 for $t > 0$, $t \notin S_1$. It is C^2 for $t > 0$, if, in addition, $y''(c_m + 0) = g''(c_m)$ which leads to relation (3.12.10) and the proof is complete.

We have just witnessed an example of proof by construction of a solution of a differential-difference equation in which the continuity of the forcing function $f(t)$ and the initial function $g(t)$ serve to characterize the existence, uniqueness and continuity properties of the solution.

In the remainder of this section we shall be concerned with the series expansion and asymptotic properties for the general system, now that we know a unique solution exists.

Consider again the differential-difference equation

$$(3.12.14) \quad \sum_{i=0}^{m} [A_i y'(t - c_i) + B_i y(t - c_i)] = f(t) \quad .$$

We define a linear operator L over the space of allowable solution vectors $y(t)$ by the equation,

$$(3.12.15) \quad L(y(t)) = \sum_{i=1}^{m} [A_i y'(t - c_i) + B_i y(t - c_i)] \quad .$$

Let us now carry out the solution of the equation $L(y) = f$ by means of the Laplace Transform. We will assume $A_0 \neq 0$, so that the solution can be shown to be exponentially bounded if $f(t)$ is. We take as the initial conditions,

$$(3.12.16) \quad y(t) = g(t), \qquad 0 \leq t \leq c_m \quad .$$

Define S, S_1 and S_2 as before and suppose g is $C^1[0, c_m]$ and f is C^0 on $([0, \infty) - S_1)$. Then $y(t)$ is C^1 for $t > 0$, $t \notin S_1$. Using the relations

$$(3.12.17) \quad \int_{c_m}^{\infty} y(t - c_i) e^{-st} dt = e^{-c_i s} \int_{c_m}^{\infty} y(t) e^{-st} dt$$

$$+ e^{-c_i s} \int_{c_m - c_i}^{c_m} g(t) e^{-st} dt \quad ,$$

and

$$(3.12.18) \quad \int_{c_m}^{\infty} y'(t - c_i) e^{-st} dt = -g(c_m - c_i) e^{-c_m s}$$

$$+ s \int_{c_m}^{\infty} y(t - c_i) e^{-st} dt \quad ,$$

we obtain

$$(3.12.19) \quad H(s) \int_{c_m}^{\infty} y(t) e^{-st} dt = P_0(s) + q(s) \quad ,$$

where

$$(3.12.20) \quad H(s) = \sum_{i=0}^{m} (A_i s + B_i) e^{-c_i s} \quad ,$$

$$(3.12.21) \quad P_0(s) = e^{-c_m s} \sum_{i=0}^{m} A_i g(c_m - c_i)$$

$$- \sum_{i=0}^{m} (A_i s + B_i) e^{-c_i s} \int_{c_m - c_i}^{c_m} g(t) e^{-st} ds \quad ,$$

$$(3.12.22) \quad q(s) = \int_{c_m}^{\infty} f(t) e^{-st} dt \quad .$$

Or, instead of using expression (3.12.17), we can use,

$$(3.12.23) \quad \int_{c_m}^{\infty} y(t - c_i) e^{-st} dt = e^{-c_i s} \int_{0}^{\infty} y(t) e^{-st} dt$$

$$- e^{-c_i s} \int_{0}^{c_m - c_i} g(t) e^{-st} dt \quad ,$$

and solve for the integral from 0 to ∞. The result is

$$(3.12.24) \quad H(s) \int_{0}^{\infty} y(t) e^{-st} dt = p(s) + q(s)$$

where,

$$(3.12.25) \quad p(s) = e^{-c_m s} \sum_{i=0}^{m} A_i g(c_m - c_i)$$

$$+ \sum_{i=0}^{m} (A_i s + B_i) e^{-c_i s} \int_{0}^{c_m - c_i} g(t) e^{-st} dt \quad .$$

With the preliminaries finished, we shall state the final result for this section. Let $y(t)$ be the continuous solution of the Eq. (3.12.14) which satisfies the initial conditions (3.12.16). Assume that $\det A_0 \neq 0$, that it is of class C^1 on $[0, c_m]$, that $f(t)$ is of class C^0 on $[0, \infty)$ except for possible jump discontinuities on the set S_1 and that $f(t)$ is exponentially bounded as $t \to \infty$. Then for any sufficiently large real number, c,

(3.12.26) $y(t) = \dfrac{1}{2\pi i} \displaystyle\int_{(c)} e^{st} H^{-1}(s)(p(s) + q(s))ds$, $t \geq 0$,

where $p(s)$ and $q(s)$ are defined in (3.12.25) and (3.12.22). If $g(t)$ is merely C^0 on $[0, c_m]$, $f(t)$ is C^0 on $[0, \infty)$ and $A_i = 0$ $(i = 1,2,\ldots,m)$, then (3.12.26) is valid for $t > c_m$, if $p(s)$ is replaced by $p_0(s)$.

Therefore by applying the Laplace Transform technique, we have been able to characterize the asymptotic behavior of the solution to the differential-difference Eq. (3.12.6) from knowledge of the characteristic function $H(s)$, the forcing function, $f(t)$ and the initial condition, $g(t)$.

3.13   A Limiting Case For Differential-Difference Equations

The differential-difference equation is related in a surprising way to a rather difficult equation arising in mathematical physics. The equation in point is the following:

(3.13.1) $\varepsilon L_m(u) + L_n(u) = 0$,

where L_m is a linear differential operator of order m, L_n is a linear differential operator of order n, $m > n \geq 1$, and ε is a small parameter. Examples of this occur in hydrodynamics where ε represents viscosity or in the study of the multivibrator (in the form of the Van der Pol equation) where ε represents a small inductance. The presence of multiplying the highest derivative in a differential equation represents a boundary layer effect which must be present to insure all boundary conditions are satisfied.

Equations of this type also arise in problems where time lags and retardations are present. If one is given the differential-difference equation of the form,

(3.13.2) $u'(t) + au(t - \varepsilon) = g(u)$, $t > \varepsilon$

$u(t) = f(t)$, $0 \leq t \leq \varepsilon$,

where $0 \leq \varepsilon \leq 1$. One frequently expands the function $u(t-\varepsilon)$ about t to obtain an approximate equation of the form,

$$(3.13.3) \quad u'(t) + au(t) - a\varepsilon u'(t) + a\varepsilon^2 u''(t)/2 = g(u) \quad ,$$

which is precisely an equation of the form given above.

We shall show that the study of the limiting behavior of the solutions of (3.13.2) is relatively easy to carry out compared to similar studies of (3.13.1). A typical result and a sketch of the proof is presented below. It is interesting to observe that we have a situation in which the more accurate description yields an easier problem, and the obvious approximation yields a more difficult problem.

Representative of the results that can be obtained from this idea is the following. Consider the equation,

$$(3.13.4) \quad u'(t) + au(t-\varepsilon) = g(u), \qquad t > \varepsilon$$
$$u(t) = f(t), \qquad 0 < t \leq \varepsilon$$

where $\varepsilon > 0$, and we require that,

(a) $g(u)$ satisfies a Lipschitz condition for $|g(t)| < c_1$

(b) $f(t)$ is continuous at $t = 0$,

(c) $|f(0)| = c_2 < c_1$,

then for $0 \leq t \leq t_0$, where t_0 is dependent upon c_1 and $g(u)$,

$$(3.13.5) \quad \lim_{\varepsilon \to 0} u(t) = v(t) \quad ,$$

where $v(t)$ is the solution of

$$(3.13.6) \quad v'(t) + av(t) = g(v), \qquad v(0) = f(0) \quad .$$

The convergence is uniform for $0 \leq t \leq t_0$.

Similar results can be obtained for systems of equations with any finite number of time lags. A brief sketch of the proof goes as follows. The proof depends, as might be expected, upon the properties of the solution to the linear equation,

(3.13.7) $u'(t) + au(t - \varepsilon) = g(t)$, $t > \varepsilon$

$$u(t) = f(t), \qquad 0 < t \leq \varepsilon ,$$

which in turn depends upon the properties of the kernel

(3.13.8) $K(t, \varepsilon) = \dfrac{1}{2\pi i} \displaystyle\int_{(C)} \dfrac{e^{st} ds}{(s + ae^{-s\varepsilon})}$.

Here (C) represents any line of the form $s = s_0 + it$, $-\infty < t < \infty$ with $s_0 > |a|$. It is easy to show that as $\varepsilon \to 0$,

(3.13.9) $K(t, \varepsilon) \to K(t) = e^{-at}$

uniformly for $0 \leq t \leq t_0 < \infty$ where t_0 is any fixed quantity.

The solution of (3.13.2), under the assumptions we have made, can be obtained as a solution of the integral equation,

(3.13.10) $u(t) = u_0(t) + \displaystyle\int_{\varepsilon}^{t} g(u(t_1)) \, K(t - t_1, \varepsilon) dt$, $t > \varepsilon$

where $u_0(t)$ is the solution of (3.13.1) corresponding to $g(u) = 0$.

We solve this equation by the method of successive approximation. Set

(3.13.11) $u_{n+1}(t) = f(t)$, $0 \leq t \leq \varepsilon$, $n \geq 0$,

$$u_{n+1}(t) = u_0(t) + \int_{\varepsilon}^{t} g(u_n(t_1)) \, K(t - t_1, \varepsilon) dt_1,$$

$$t > \varepsilon, \qquad n \geq 0 .$$

Under the foregoing assumptions, it is easy to show that $u_n(t) \to u(t)$ uniformly for $0 \leq \varepsilon \leq \varepsilon_1$, $0 \leq t \leq t_0$, for $n = 0, 1, \ldots$ and that as $\varepsilon \to 0$, $[u_n(t) - v_n(t)] \to 0$ uniformly for $0 \leq t \leq t_0$, where the sequence $v_n(t)$ is determined by the relation

$$(3.13.12) \qquad v_0(t) = f(t)$$

$$v_{n+1}(t) = v_0(t) + \int_0^t g(v_n(t_1))e^{-a(t-t_1)} dt_1 ,$$

$$t \geq 0, \quad n = 0,1,2,\ldots$$

Since the same assumptions ensure that $v_n(t) \to v(t)$ uniformly in $0 \leq t \leq t_0$, where $v(t)$ is the solution of,

$$(3.13.13) \quad v(t) = f(0) + \int_0^t g(v(t_1))e^{-a(t-t_1)} dt_1 ,$$

we have the desired uniform convergence of $u(t)$ to $v(t)$ on $0 \leq t \leq t_0$, as $\varepsilon \to 0$.

The same method can be applied to demonstrate the corresponding result for more general differential-difference equations.

Problems

1. Use the continuation process to calculate the solution of
$$u'(t) = 1 + u(t-1) ,$$
$$u(t) = 1, \qquad 0 < t < 1 ,$$
in the interval $n < t < n+1$.

2. Use the continuation process to calculate the solution of
$$u'(t) = 2u(t-1) , \qquad t > 1$$
$$u(t) = t, \qquad 0 < t < 1$$
for $0 < t < 5$.

3. Show that $u = e^{st}$ is a solution of $u'(t) = u(t-1)$ if s is a solution of the transcendental equation $s = e^{-s}$. What is the corresponding initial condition for u in this problem?

4. Use the above procedure to find the solution of the linear first-order differential equation

$$a_0 u'(t) + b_0 u(t) = 0, \qquad t > 0, \qquad u(0) = u_0$$

in the form

$$u(t) = \frac{1}{2\pi i} \int_{(c)} \frac{a_0 u_0}{a_0 s + b_0} e^{st} ds, \qquad t > 0, \qquad c > - b_0/a_0 .$$

5. Deduce from Problem 1 that

$$a_0 \int_{(c)} \frac{e^{st}}{a_0 s + b_0} ds = \exp(- b_0 t/a_0), \qquad t > 0 ,$$

for any $c > - b_0/a_0$.

6. Using the Laplace Transform method, show that the solution of

$$a_0 u'(t) + b_0 u(t) = f(t) , \qquad t > 0 ,$$

has the form

$$u(t) = \int_{(c)} \frac{a_0 u_0 + \int_0^\infty f(t_1) e^{-st_1} dt_1}{a_0 s + b_0} e^{st} ds , \qquad t > 0 .$$

7. For the equation

$$u'(t) - u(t-1) = 1 , \qquad t > 1$$
$$u(t) = 1 , \qquad 0 \le t \le 1 ,$$

show that

$$\int_0^\infty u(t) e^{-st} dt = \frac{s + e^{-s}}{s(s - e^{-s})} .$$

Hence,

$$u(t) = -1 + 2 \sum_{n=0}^{N} (t-n)^n/n! , \qquad N \le t \le N+1 ,$$
$$N = 0,1,2,\ldots$$

8. For the equation

$$u'(t) = 2u(t-1), \qquad t > 1$$
$$u(t) = t, \qquad 0 \le t \le 1,$$

show that

$$u(t) = \frac{1}{2} + \sum_{n=1}^{N} \left[\frac{2^n (t-n)^{n+1}}{(n+1)!} - \frac{2^{n-1}(t-n)^n}{n!} \right]$$

for $N \le t \le N+1$; $\quad N = 0,1,\ldots$

9. Let $u(t)$ be the continuous solution of

$$u'(t) + b_1 u(t-c) = 0, \qquad t > c$$
$$u(t) = g(t), \qquad 0 \le t \le c.$$

Define $g_1(t) = g'(t)H(c-t), \quad t > 0$

$$g_{-n}(t) = \int_0^t g_{1-n}(t_1)dt_1, \qquad t > 0, \qquad n = 0,1,2,\ldots$$

Show that for $t > 0$,

$$u(t) = g(0) \sum_{n=0}^{\infty} (-b_1)^n / n! (t-nc)^n H(t-nc)$$

$$+ \sum_{n=0}^{\infty} (-b_1)^n g_{-n}(t-nc) H(t-nc).$$

Hint: Use the expansion of $h^{-1}(s)$ in powers of $s^{-1}e^{-cs}$, and show that

$$s^{-n-1} \int_0^c g'(t_1)e^{-st_1} dt_1$$

is the transform of $g_{-n}(t)$.

Chapter 4

PARTIAL DIFFERENTIAL EQUATIONS

4.1   Introduction

The application of the Laplace Transform technique to the solution of partial differential equations has proven to be very useful. When an analytic representation of the solution can be obtained, valuable insight can be found about the behavior of transient solutions as well as asymptotic trends in the solution. With the use of the large scale computer currently in use today, a numerical solution to a partial differential equation may be found by first reducing the partial to an ordinary differential equation by means of the Laplace Transform, solving the resulting equation numerically as a function of the complex parameter s and finally performing a numerical inversion of the final Laplace Transform. These techniques will be discussed in a later chapter.

A partial differential equation is an equation governing the differential behavior of a function $f(x_1, x_2,...,x_n)$ over a clearly defined region in the n-dimensional space of the set of independent variables $x_1, x_2,...,x_n$. Thus,

(4.1.1) $F(x_1, x_2,...,x_n, f, \partial f/\partial x_1,...,\partial^p f/\partial x_n^p) = 0$

represents a partial differential equation governing the behavior of the function $f(x_1, x_2,...,x_n)$ over a region defined by the surface $g(x_1, x_2,...,x_n) = 0$. The dimension of the system (4.1.1)

is defined by the positive integer n and the order of the
equation is given by the positive integer p.

The use of the Laplace Transform in solving problems
involving partial differential equation has potentially a great
application when the equation under study is linear.

Since the emphasis of this book is with the Laplace Transform, we shall consider only those problems where one of the
independent variables is varied from zero to infinity. This
restriction still allows us great flexibility for it includes all
problems dealing with dynamic behavior $(t \geq 0)$ and among others,
static problems where one or more dimensions are semi-infinite.

4.2   The Hyperbolic Partial Differential Equation

Consider the following problem. One wishes to impose a
voltage at one end of a straight conducting wire of length ℓ at
time $t = 0$. We ask, what is the subsequent history of voltage at
any location in the wire at any time? Let e = volts, i = current,
and x the length of the wire.

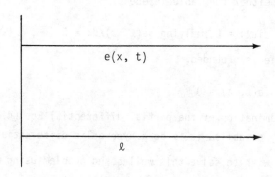

$e(x, t)$

ℓ

Fig. 4.1

The parameters of the system are L, the inductance per unit length of cable (henries/ℓ) and C the capacitance to ground per length, (farads/ℓ). The mathematical model for the system is,

(4.2.1) $\partial e/\partial x = -L\partial i/\partial t$; $\partial i/\partial x = -C\partial e/\partial t$.

Combining the two equations given in (4.2.1) we can describe the system either through the voltage e or in the current i. Thus,

(4.2.2) $\partial^2 e/\partial x^2 = LC\partial^2 e/\partial t^2$,

or

(4.2.3) $\partial^2 i/\partial x^2 = LC\partial^2 i/\partial t^2$.

Initially the cable is dead,

(4.2.4) $e(x, 0) = \partial e(x, 0)/\partial t = 0$.

For boundary conditions, the following conditions are imposed, at $t = 0$, a voltage is impressed at $x = 0$,

(4.2.5) $e(0, t) = f(t)H(t)$,

at $x = \ell$, either the cable is open,

(4.2.6) $i(\ell) = 0$ implying $de(\ell, t)/dx = 0$,

or the cable is grounded,

(4.2.7) $e(\ell, t) = 0$.

From an examination of the partial differential Eq. (4.2.2) we can quickly establish it to be a hyperbolic differential equation.

        We wish to solve this well posed problem using the Laplace Transform techniques. We start by defining the Laplace Transform of the function $e(x, s)$ as,

(4.2.8) $e(x, s) = \int_0^\infty e^{-st} e(x, t)dt = L(e(x, t))$.

Several properties of the function $L(e(x, t))$ must be stated.

$$(4.2.9) \qquad L(\partial e(x, t)/\partial x) = de(x, s)/dx$$

$$L(\partial e(x, t)/\partial t) = se(x, s) - e(x, 0)$$

$$L(\partial^2 e(x, t)/\partial t^2) = s^2 e(x,s) - se(x,0) - \partial e(x,0) /\partial t \quad .$$

If we multiply (4.2.2) by e^{-st}, integrate the resultant equation from zero to infinity and apply the results given in (4.2.9), then we have,

$$(4.2.10) \qquad d^2 e(x,s) /dx^2 = LC(s^2 e(x, s) - se(x, 0) - \partial e(x, 0)/\partial t) \quad .$$

In examining (4.2.10) we note several important points. First, in forming the transform variable $e(x, s)$, the variable t is integrated out and replaced by the parameter s and second the initial conditions which appeared as auxiliary conditions for the partial differential Eq. (4.2.2) now are part of the forcing function in the ordinary differential Eq. (4.2.10). Recalling the initial conditions of the problem, (4.2.4), the equation which must be solved is,

$$(4.2.11) \qquad d^2 e(x, s)/dx^2 - \left(\frac{s}{v}\right)^2 e(x, s) = 0 \qquad ,$$

where $v^2 = 1/LC$, having the dimension of velocity squared. In terms of the parameter s, the solution to (4.2.11) is,

$$(4.2.12) \qquad e(x, s) = A(s)e^{-sx/v} + B(s)e^{sx/v} \qquad .$$

The two arbitrary functions $A(s)$ and $B(s)$ can be found from the imposed boundary conditions. Recalling that at $x = 0$, $e(0, t) = f(t)$, $t \geq 0$, the Laplace Transform of $f(t)$ is,

$$(4.2.13) \qquad F(s) = \int_0^\infty e^{-st} f(t)dt \qquad .$$

Hence the two boundary conditions for the cable are,

(4.2.14) $A(s) + B(s) = F(s)$

(4.2.15) $- A(s)e^{-s\ell/v} + B(s)e^{s\ell/v} = 0$.

Equation (4.2.15) is derived from the boundary condition $de(\ell, s)/dx = 0$. Solving (2.4.14) and (2.4.15) for $A(s)$ and $B(s)$ we get,

(4.2.16) $A(s) = \left\{ \dfrac{F(s)}{(1 + e^{-2s\ell/v})} \right\}$

(4.2.17) $B(s) = \left\{ \dfrac{F(s)e^{-2s\ell/v}}{(1 + e^{-2s\ell/v})} \right\}$.

If the cable is of infinite length, $\ell \to \infty$, then

$A(s) = F(s)$ and $B(s) = 0$.

Thus, the solution to the ordinary differential equation in the transform variable is

(4.2.18) $e(x, s) = F(s)e^{-sx/v}$.

Now, if we apply the Laplace inversion formula to this problem, we get,

(4.2.19) $e(x, t) = \dfrac{1}{2\pi i} \displaystyle\int_{-\infty i+c}^{\infty i+c} e^{st} F(s)e^{-(sx/v)} ds$

$= \dfrac{1}{2\pi i} \displaystyle\int_{-\infty i+c}^{\infty i+c} e^{(t - x/v)s} F(s) ds$

$= \begin{cases} f(t - x/v) & t > x/v \\ 0 & t < x/v \end{cases}$.

Hence for an infinitely long cable, the wave form imposed in the transmission line at $x = 0$, is propagated without distortion along the cable at a propagation velocity of $\ell/\sqrt{LC}$.

If the cable is of finite length, then the solution in the transformed variable s is, using expressions (4.2.16) and (4.2.17) in (4.2.12),

$$(4.2.20) \quad e(x, s) = F(s)/(1 + e^{-2s\ell/v})(e^{-sx/v} + e^{-s(2\ell - x)/v}) .$$

To put $e(x, s)$ into a more useful form, we expand

$$(4.2.21) \quad 1/(1 + e^{-2s\ell/v}) = 1 + e^{-2s\ell/v} + e^{-4s\ell/v} + \dots$$

Hence,

$$(4.2.22) \quad e(x, s) = F(s)\left[e^{-sx/v} - e^{-s(2\ell - x)/v} + e^{-s(2\ell + x)/v}\right.$$
$$\left. - e^{-s(4\ell - x)/v} + e^{-s(4\ell + x)/v} + \dots\right]$$

If we perform the inverse of the Laplace Transform we get,

$$e(x, t) = \begin{cases} f(t - x/v) & t > x/v \\ 0 & t < x/v \end{cases}$$

$$+ \begin{cases} f(t - (2\ell + x)/v) & t > (2\ell + x)/v \\ 0 & t < (2\ell + x)/v \end{cases}$$

$$- \begin{cases} f(t - (2\ell - x)/v) & t > (2\ell - x)/v \\ 0 & t < (2\ell - x)/v \end{cases}$$

$$- \begin{cases} f(t - (4\ell - x)/v) & t > (4\ell - x)/v \\ 0 & t < (4\ell - x)/v \end{cases}$$

$$+ \begin{cases} f(t - (4\ell + x)/v) & t > (4\ell + x)/v \\ 0 & t < (4\ell + x)/v \end{cases}$$

$$+ \quad \dots$$

The solution for the finite cable reflects clearly the propagation of the signal down the cable followed by signal distortion caused by the superposition of reflected wave forms as the signal reacts to the end condition. Hence, we have seen in this example, the straightforward power of the Laplace technique in solving partial differential equations of the hyperbolic type. As we shall now see, the procedure is even more interesting when we consider partial differential equations of the parabolic and elliptic type, as we shall do in the next two sections.

4.3   The Parabolic Partial Differential Equation

The parabolic partial differential equation can be used as a model for describing the one dimensional flow of heat through a uniform medium when the initial condition and boundary condition are known. We can apply the Laplace Transform technique to solving equations of this form and in this section we will show how this is done. The linear parabolic partial differential equation takes the form,

$$(4.3.1) \qquad \frac{\partial^2}{\partial x^2} w(x, t) = \left(\frac{1}{c^2}\right) \partial w(x, t)/dt$$

where the initial and boundary conditions are generally defined as,

$$(4.3.2) \qquad w(x, 0) = f(x) \qquad 0 \leq x \leq L$$

$$w(0, t) = g_0(t) \qquad t \geq 0$$

$$w(L, t) = g_L(t) \qquad .$$

We wish to solve the partial differential Eq. (4.3.1), subject to the initial and boundary conditions given by (4.3.2), by the method of the Laplace Transform.

As usual we introduce the function,

$$(4.3.3) \qquad w(x, s) = \int_0^\infty e^{-st} w(x, t) dt \qquad .$$

With this definition, Eq. (4.3.1) can be transformed into an ordinary differential equation by multiplying (4.3.1) by e^{-st} and integrating the results from zero to infinity. By integrating by parts and applying the initial conditions, we have,

$$(4.3.4) \qquad d^2w(x, s)/dx^2 - s/c^2 w(x, s) = -f(x)/c^2 \quad .$$

If we first consider the homogeneous equation where $f(x) = 0$, the solution to the homogeneous equation can be written down immediately as,

$$(4.3.5) \qquad w(x, s) = A(s)e^{(-\sqrt{s}/c)x} + B(s)e^{(\sqrt{s}/c)x}$$

where the functions $A(s)$ and $B(s)$ will eventually be determined by the boundary conditions.

Before we can do this, however, we must determine the particular solution to (4.3.1) which reflects the initial conditions to the problem and we choose to apply the variation of parameters technique. We begin by observing that the two solutions to the homogeneous equation are,

$$(4.3.6) \qquad u_1(x) = e^{(-\sqrt{s}/c)x}, \qquad u_2(x) = e^{(\sqrt{s}/c)x} \quad .$$

We will now consider a solution to (4.3.4) of the form,

$$(4.3.7) \qquad w(x, s) = a(x)u_1(x) + b(x)u_2(x) \quad ,$$

where the functions $a(x)$ and $b(x)$ are unknown functions to be determined.

We differentiate (4.3.7)

$$(4.3.8) \qquad w'(x, s) = a(x)u_1'(x) + a'(x)u_1(x) + b(x)u_2'(x) + b'(x)u_2(x)$$

and require that,

$$(4.3.9) \qquad a'(x)u_1(x) + b'(x)u_2(x) = 0 \quad .$$

Differentiating (4.3.8) again, using (4.3.9), we have,

$$(4.3.10) \qquad w''(x, s) = a(x)u_1''(x) + a'(x)u_1'(x) + b(x)u_2''(x)$$
$$+ b'(x)u_2'(x) \quad .$$

Substituting (4.3.10) and (4.3.6) into (4.3.4), and noting that u_1 and u_2 are solutions to the homogeneous equation, we are left with two equations defining the unknown functions $a(x)$ and $b(x)$.

$$(4.3.11) \qquad (da/dx)(du_1/dx) + (db/dx)(du_2/dx) = -f(x)/c^2$$
$$da/dxu_1 + db/dxu_2 = 0 \quad .$$

Solving this system we find,

$$(4.3.12) \qquad db/dx = -(f(x)/c^2)(u_1/(u_1du_2/dx - u_2du_1/dx))$$
$$da/dx = (f(x)/c^2)(u_2/(u_1du_2/dx - u_2du_1/dx)) \quad .$$

And by integrating from zero to x, we get,

$$(4.3.13) \qquad a(x) = (1/c^2)\int_0^x f(r)(u_2(r)/(u_1(r)du_2(r)/dr$$
$$- u_2(r)du_1(r)/dr))dr$$

$$(4.3.14) \qquad b(x) = - (1/c^2)\int_0^x f(r)(u_1(r)/(u_1(r)du_2(r)/dr$$
$$- u_2(r)du_1(r)/dr))dr$$

and the particular solution to (4.3.4) is,

$$(4.3.15) \qquad W_p(x, s) = a(x)u_1(x) + b(x)u_2(x) \quad .$$

Using the specific form of $u_1(x)$ and $u_2(x)$ given in (4.3.6), we can write the solution to (4.3.4) as,

(4.3.16) $\quad w(x, s) = Ae^{-(\sqrt{s}/c)x} + Be^{(\sqrt{s}/c)x}$

$$+ (1/c\sqrt{s}) \int_0^x f(r)\sinh((\sqrt{s}/c)(x - r))dr \quad .$$

With the complete solution of (4.3.4) given in (4.3.16) we are now in a position to apply the two time dependent boundary conditions. Let $g_0(s)$ and $g_L(s)$ be the Laplace Transform of $g_0(t)$ and $g_L(t)$ respectively. The boundary conditions require that

(4.3.17) $\quad A + B = g_0(s)$

$$Ae^{-(\sqrt{s}/c)L} + Be^{(\sqrt{s}/c)L}$$

$$+ (1/c\sqrt{s}) \int_0^L f(r)\sinh((\sqrt{s}/c)(L - r))dr = g_L(s) \quad .$$

Solving for the unknown constants, we have a solution to (4.3.4) in the form,

(4.3.18)

$$A = \frac{(g_L(s) - g_0(s)e^{(\sqrt{s}/c)L}) + (1/c\sqrt{s}) \int_0^L f(r)\sinh((\sqrt{s}/c)(L - r))dr}{-2 \sinh((\sqrt{s}/c)L)}$$

(4.3.19)

$$B = \frac{-(-g_L(s) + g_0(s)e^{-(\sqrt{s}/c)L}) - (1/c\sqrt{s}) \int_0^L f(r)\sinh((\sqrt{s}/c)(L - r))dr}{-2 \sinh((\sqrt{s}/c)L)} ,$$

(4.3.20) $\quad w(x, s) = Ae^{-(\sqrt{s}/c)x} + Be^{(\sqrt{s}/c)x}$

$$+ (1/c\sqrt{s}) \int_0^x f(r)\sinh((\sqrt{s}/c)(L - r))dr \quad .$$

Once we have a complete definition of the solution to the inhomogeneous differential equation (4.3.4), subject to the boundary conditions defined in (4.3.17), the remaining task is to evaluate the Laplace inversion formula,

$$(4.3.21) \qquad w(x, t) = \frac{1}{2\pi i} \int_{(c)} e^{st} w(x, s) ds$$

by the theory of residues.

Even before the boundary conditions are completely defined, the form of the solution (4.3.20) gives some indication as to where the poles could occur in the complex s plane. For example, using the fact that,

$$(4.3.22) \qquad \sinh(ix) = i \sin(x)$$

at first glance, poles of (4.3.20) could occur when,

$$(4.3.23) \qquad (s/c)L = in\pi$$

or

$$s = -c^2 n^2 \pi^2 / L^2 \qquad ,$$

i.e., poles could lie along the negative real axis in the complex s-plane. The presence of the $\sqrt{s}$ in the denominator of (4.3.20) suggests that a singularity might exist at $s = 0$, yet a careful check of this possibility proves otherwise. To proceed with the analysis, it is necessary to define both the initial condition and the boundary conditions in a specific way. As a simple example, the initial condition to be,

$$(4.3.24) \qquad f(x) = \cos(\pi x/L) \qquad ,$$

while the two boundary conditions are

$$(4.3.25) \qquad g_0(t) = e^{-(c^2 \pi^2 / L^2)t},$$

$$g_L(t) = -e^{-(c^2 \pi^2 / L^2)t} \qquad .$$

We wish to demonstrate the use of the Laplace inverse formula (4.3.21), when both the initial condition and boundary conditions are given. Taking the Laplace Transform of the boundary conditions (4.3.25), we can write,

$$(4.3.26) \qquad g_0(s) = 1/(s + c^2\pi^2/L^2)$$

$$g_L(s) = -1/(s + c^2\pi^2/L^2) \qquad .$$

From the initial condition, $f(x) = \cos(\pi x/L)$, we can directly evaluate,

$$(4.3.27) \qquad (1/c\sqrt{s})\int_0^x \cos(\pi r/L) \sinh((\sqrt{s}/c)(x-r)dr$$

$$= 1/(s + c^2\pi^2/L^2)[-\cos(\pi x/L) + \cosh((\sqrt{s}/c)x)] \qquad .$$

Substituting the results given in (4.3.26) and (4.3.27) into (4.3.20) and after a bit of manipulation, we have,

$$(4.3.28) \qquad w(x, s) = \frac{2\cosh((\sqrt{s}/c)x) - \cos(\pi x/L)}{(s + c^2\pi^2/L^2)} \qquad .$$

By looking at (4.3.28) it is immediately clear that the suspected poles which were found to lie along the negative real axis in the complex s-plane, are cancelled due to the specific choice of the boundary and initial conditions leaving only a single simple pole at $s = -c^2\pi^2/L^2$. Evaluating the solution, given in (4.3.21) by the theory of residues, we have

$$(4.3.29) \qquad w(x, t) = e^{-(c^2\pi^2/L^2)t}\cos(\pi x/L) \qquad .$$

It is instructive to note, that in problems of this kind, the final placement of the poles of the complex function $w(x, s)$ depend not only on the form of the partial differential equation but also on the precise form of both the initial and boundary conditions.

4.4   The Elliptic Partial Differential Equation

The elliptic partial differential equation is closely

associated with such problems as the vibration of strings or thin
membranes. Typically such an equation has the form,

(4.4.1) $\partial^2 z(x, t)/\partial x^2 + (1/a^2)\partial^2 z(x, t)/\partial t^2 = 0,$ $0 \leq x \leq L,$

$t \geq 0$.

Taking the Laplace Transform of (4.4.1) gives,

(4.4.2) $d^2 z(x, s)/dx^2 + s^2/a^2 z(x, s) - \dot{z}(x, 0) - s z(x, 0) = 0$.

Initially, if we let the system begin with zero displacement and
velocity, i.e., $z(x, 0) = 0$ and $\dot{z}(x, 0) = 0,$ then (4.4.2)
becomes,

(4.4.3) $d^2 z(x, s)/dx^2 + (s/a)^2 z(x, s) = 0$

whose solution immediately can be written as,

(4.4.4) $z(x, s) = A(s)\sin(sx/a) + B(s)\cos(sx/a)$.

If we define the boundary conditions associated with (4.4.1) to be,

(4.4.5) $z(0, t) = g_0(t)$

$z(L, t) = g_L(t)$,

and if the Laplace Transform of both functions are $g_0(s),$ and
$g_L(s),$ then the boundary conditions given in (4.4.5) require that,

(4.4.6) $B(s) = g_0(s)$,

$A(s) = (g_L(s) - g_0(s)\cos(sL/a))/\sin(sL/a)$.

The solution in the transformed variable s is,

(4.4.7) $z(x, s) = (g_L(s) - g_0(s)\cos(sL/a))(\sin(sx/a)/\sin(sL/a))$

$+ g_0(s)\cos(sx/a)$.

In order to proceed further we must select a precise form of the two boundary conditions. Therefore let us define,

(4.4.8) $\quad g_0(t) = a_1 + b_1 t$

$\quad\quad\quad g_L(t) = a_2 + b_2 t$

whose Laplace Transforms are,

(4.4.9) $\quad g_0(s) = a_1/s + b_1/s^2$

$\quad\quad\quad g_L(s) = a_2/s + b_2/s^2 \quad .$

The solution now becomes

(4.4.10)

$$z(x, s) = ((a_2/s + b_2/s^2)$$

$$- (a_1/s + b_1/s^2)(\cos(sL/a))(\sin(sx/a)/\sin(sL/a)$$

$$+ (a_1/s + b_1/s^2)\cos(sx/a)) \quad .$$

If we recall from Chapter 2,

(4.4.11) $\quad z(x, t) = \dfrac{1}{2\pi i} \displaystyle\int_{-\infty i + c}^{\infty i + c} e^{st}\, z(x, s)\, ds$

and

$$\int_C e^{st}\, z(x, s)\, ds = 2\pi i \sum_{j=1}^{\infty} K_j \quad ,$$

where K_j are the residues of the poles at z_j, within the closed contour C in the complex plane. If we recall from the definition of the residues, if $f(z)$ has a pole of order m at $z = a$, then,

(4.4.12) $\quad K_j = (1/(m-1)!)\, d^{m-1}/dz^{m-1}((z-a)^m\, f(z))_{z=a} \quad .$

Therefore if we identify the poles of the integrand of (4.4.11) and determine the residues at each pole, then for an appropriately defined closed contour, the form of $z(x, t)$ is the sum of the residues within the closed contour C.

A close examination of the poles of (4.4.10), show that poles occur at,

(4.4.13) $s = 0$

$s = n\pi/L$, $n = 0,1,2,...$

If we rewrite the transform solution (4.4.11) in the form,

(4.4.14) $z(x, s) = \dfrac{(a_2 - a_1\cos(sL))}{s\left(\dfrac{\sin sL}{\sin sx}\right)} + \dfrac{(b_2 - b_1\cos(sL))}{s^2\left(\dfrac{\sin sL}{\sin sx}\right)}$

$+ \left(\dfrac{a_1}{s} + \dfrac{b_1}{s^2}\right)\cos(sx)$.

The first term in (4.4.14) has a simple pole $(m = 1)$ at $s = 0$, the residue associated with this pole can be shown to be,

(4.4.15) $K_1 = (a_2 - a_1)x/L$.

The second term in (4.4.14) has a second-order pole $(m = 2)$ at $s = 0$, with a residue of,

(4.4.16) $K_2 = (b_2 - b_1)xt/L$.

The third term in (4.4.14) yields a residue of

(4.4.17) $K_3 = a_1$

while the fourth term in (4.4.14) although having an order of $2(m = 2)$ at the origin $(s = 0)$, has a residue of zero.

(4.4.18) $K_4 = bt$.

The complete solution to (4.4.1) is therefore,

(4.4.19) $z(x, t) = a_1 + (a_2 - a_1)x/L + b_1 t + (b_2 - b_1)xt/L$.

Clearly this satisfies the partial differential Eq. (4.4.1). As for the boundary conditions,

(4.4.20) $z(0, t) = a_1 + b_1 t$

$z(L, t) = a_2 + b_2 t$,

while the initial condition is,

(4.4.21) $z(x, 0) = a_1 + (a_2 - a_1)x/L$,   which is zero,

if $a_1 = a_2 = 0$.

Chapter 5

THE RENEWAL EQUATION

5.1   Introduction

The solution of an ordinary differential equation by means of the Laplace Transform was presented in Chapter 2. Another form of equation which is equally suited to solution by the transform method is the integral equation which we will be concerned with throughout this chapter. As an example we will consider the alternative form of a differential equation as an integral equation and proceed to solve this equation by the Laplace Transform technique.

With this as an introduction, we will consider some of the important properties of the derived integral equation, which is, in fact, the renewal equation.

Let us consider the equation for a vibrating string (harmonic oscillator).

(5.1.1) $d^2f(t)/dt^2 + k^2f(t) = 0$

$$f(0) = f_0$$

$$df(0)/dt = v_0 \quad .$$

At this point in the analysis we introduce the Green's function G which is a function of two variables t and τ and satisfying the differential equation,

$$(5.1.2) \qquad \frac{d^2 G(\tau,t)}{dt^2} = - \delta(\tau - t)$$

$$G(\tau,t) = 0 , \qquad t \leq \tau$$

where $\delta(\tau - t)$ is the Dirac Delta or impulse function, i.e.,

$$\delta(t) = 0, \qquad t \neq 0 \qquad \text{and} \qquad \int_\infty^\infty \delta(t)dt = 1 .$$

If we multiply Eq. (5.1.1) by G, Eq. (5.1.2) by f, subtract the two and integrate the results from zero to t, we have,

$$(5.1.3) \qquad \int_0^t \Big(G(d^2f/dt^2 + k^2 f) - f(d^2 G/dt^2 + \delta(\tau - t))\Big) dt = 0 .$$

This can be rewritten as,

$$(5.1.4) \qquad \int_0^t (Gd^2f/dt^2 - fd^2G/dt^2)dt + k^2\int_0^t Gfdt = f(\tau) .$$

The first integral can be rewritten by integrating by parts, giving the result,

$$(5.1.5) \qquad f_0\, dG(t,0)/dt - G(t,0)v_0 + k^2\int_0^t G(t,\tau)f(\tau)d\tau = f(t) .$$

From Eq. (5.1.2), we can immediately write down the solution for the Green's function G,

$$(5.1.6) \qquad G(t,\tau) = (\tau - t)H(t - \tau)$$

where $H(t - \tau)$ is the unit step function.

$$H(t) = \begin{cases} 1 & t > 0 \\ 0 & t < 0 \end{cases} .$$

Putting this form of G into Eq. (5.1.5) gives

$$(5.1.7) \qquad f_0 + v_0 t + k^2\int_0^t (\tau - t)f(\tau)d\tau = f(t) .$$

Equation (5.1.7) is the integral equation form of the differential equation given in (5.1.1). In this form the initial conditions are incorporated directly into the equation rather than as auxiliary conditions as they appear in (5.1.1). We propose to solve Eq. (5.1.7) by the Laplace Transform techniques. In doing so we will demonstrate the technique as well as show that the solution agrees with that found in Chapter 2.

Taking the Laplace Transform of (5.1.7), we get

$$(5.1.8) \qquad f(s) = f_0/s + v_0/s^2 - k^2 \int_0^\infty e^{-st} \int_0^t (\tau - t)f(\tau)d\tau \quad .$$

By applying the convolution theorem given in Chapter 1, we have

$$(5.1.9) \qquad L\left(\int_0^x v(x - x_0)f(x_0)dx_0\right) = v(s)f(s) \quad ,$$

which when placed in (5.1.8) gives,

$$(5.1.10) \qquad f(s) = f_0/s + v_0/s^2 - k^2 f(s)/s^2 \quad .$$

After solving for $f(s)$,

$$(5.1.11) \qquad f(s) = (v_0 + sf_0)/(s^2 + k^2) \quad ,$$

we can write the solution as the inverse Laplace Transform. Thus

$$(5.1.12) \qquad f(t) = \frac{1}{2\pi i} \int_{-\infty i+c}^{\infty i+c} (v_0 + sf_0)/(s^2 + k^2)e^{st}ds \quad .$$

The contour taken to evaluate (5.1.12) is shown in Fig. 5.1.

In the evaluation of (5.1.12) by the contour given in Fig. 5.1, the contribution of the semicircular path goes to zero as the contour approaches infinity. In such a closed contour, the only contribution comes from the poles located at $+ik$ and $-ik$ on the imaginary axis.

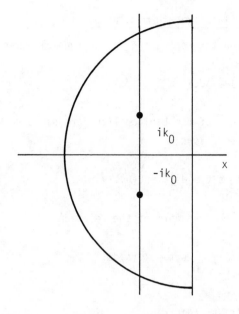

Fig. 5.1

Rewriting (5.1.12) we see that

$$(5.1.13) \qquad f(t) = \frac{1}{2\pi i} \int_{-\infty i + c}^{\infty i + c} \frac{(v_0 + sf_0)e^{st}}{2ik} \left(\frac{1}{(s-ik)} - \frac{1}{(s+ik)} \right) ds \quad .$$

When the contour integral is evaluated at the poles $+ik$ and $-ik$, the solution is found to be,

$$(5.1.14) \qquad f(t) = f_0 \sin(kt) + (v_0/k)\cos(kt) \quad .$$

The form of the integral Eq. (5.1.7) is

$$(5.1.15) \qquad \phi(t) + \int_0^t g(t-\tau)f(\tau)d\tau = f(t) \quad .$$

This linear functional equation is known as the renewal equation which we shall consider in detail throughout this chapter.

5.2   The Formal Laplace Transform Solution

The convolution term appearing in the renewal equation,

$$(5.2.1) \qquad \int_0^t u(t-s)v(s)ds$$

immediately suggests use of the Laplace Transform since we know that under suitable conditions,

$$(5.2.2) \qquad L\left(\int_0^t u(t-s)v(s)ds\right) = L(u)L(v) \quad .$$

Hence taking the Laplace Transform of the relation

$$(5.2.3) \qquad u(t) = f(t) + \int_0^t u(t-s)v(s)ds \qquad ,$$

we have

$$(5.2.4) \qquad L(u) = L(f) + L(u)L(v)$$

or

$$(5.2.5) \qquad L(u) = \frac{L(f)}{1 - L(v)} \quad .$$

Hence we suggest that with appropriate conditions imposed on f and v, we will have the explicit solution,

$$(5.2.6) \qquad u(t) = \frac{1}{2\pi i} \int_{(c)} \frac{e^{st} \int_0^\infty e^{-sp}f(p)dp}{\left(1 - \int_0^\infty e^{-sp}v(p)dp\right)} ds \quad .$$

5.3   Exponential Bounds on u(t)

In this section we shall derive some simple conditions which permit us to use the Laplace Transform technique to solve the renewal equation.

Let us establish the following result.

If for $t \geq 0$ and some a, we have

(5.3.1)   (a) $|f(t)| \leq c_1 e^{at}$

(b) $\displaystyle\int_0^\infty e^{-at} |v(t)| dt = c_2 < 1$

then

(5.3.2) $|u(t)| \leq \dfrac{c_1 e^{at}}{1 - c_2}$.

The point of this result is that $u(t)$ is bounded by an exponential whenever $f(t)$ and $v(t)$ are. This condition is met in all important applications. It follows that $L(u)$ will be an analytic function of s for $Re(s)$ sufficiently large.

To establish this we proceed as follows. We have from the renewal equation and condition (a)

(5.3.3) $|u| \leq c_1 e^{at} + \displaystyle\int_0^t |u(t-s)||v(s)| ds$.

Hence

(5.3.4) $|u e^{-at}| \leq c_1 + \displaystyle\int_0^t |u(t-s) e^{-a(t-s)}| \; |e^{-as} v(s)| ds$.

Write

(5.3.5) $w(t) = \displaystyle\max_{0 \leq t_1 \leq t} |u(t_1) e^{-at_1}|$.

Then (5.3.4) yields,

(5.3.6) $w(t) \leq c_1 + w(t) \displaystyle\int_0^t e^{-as} |v(s)| ds$

$\leq c_1 + w(t) \displaystyle\int_0^\infty e^{-as} |v(s)| ds \leq c_1 + c_2 v(t)$.

From this we obtain (5.3.2).

Using the foregoing result, we readily establish the following.

If for some a we have,

(5.3.7)　　(a)　$|f(t)| \leq c_1 e^{at}$,　　$t \geq 0$,

　　　　　　(b)　$\int_0^\infty e^{-at} |v(t)| dt < 1$,

then the Laplace Transform of u, L(u), is given by

(5.3.8)　　$L(u) = \dfrac{L(f)}{1 - L(v)}$

for $Re(s) > a$.

Hence at every point t where u is continuous and of bounded variation in some interval containing t, we have

(5.3.9)　　$u = \dfrac{1}{2\pi i} \int_{(b)} \dfrac{L(f)}{1 - L(v)} e^{st} ds$

for $b > a$.

In the previous section, we gave simple conditions upon f and v which permitted us to conclude that u is continuous and is of bounded variation.

5.4　A Convolution Theorem

The solution of

(5.4.1)　　$w(t) = f(t) + \int_0^t w(t-s) v(s) ds$

is a certain linear operation on f(t). It is of interest to determine the precise form of this operation and it turns out that this operation is obtained from the solution of the simpler equation

(5.4.2)　　$u(t) = 1 + \int_0^t u(t-s) v(s) ds$.

This result can be of service in connection with the study of the asymptotic behavior of the solution. In order to derive the formula, let us use the Laplace Transform in a heuristic fashion. We have, by our previous results

(5.4.3) $L(u) = 1/(s(1 - L(v)))$

$L(w) = L(f)/(1 - L(v))$

whence

(5.4.4) $L(w)/L(u) = sL(f) = f(0) + \int_0^a e^{-st} f'(t)dt$.

From the convolution theorem it follows that

(5.4.5) $w(t) = f(0)u(t) + \int_0^t u(t - s)f'(s)ds$.

It is this formula which we wish to establish rigorously, under appropriate assumptions on f and v. If

(5.4.6)   (a) $f'(t)$ exists for $0 \le t \le t_0$, $\int_0^{t_0} |f'(t)|ds < \infty$

and

(b) $\int_0^{t_0} |v(s)|ds < \infty$,

then a solution to (5.4.1) is given by (5.4.5) for $0 \le t \le t_0$.

To show this result, let u be the solution of the equation in (5.4.2) and let w be the function defined in (5.4.5). Then we have

(5.4.7) $\int_0^t w(t - s)v(s)ds = f(0)\int_0^t u(t - s)v(s)ds$

$+ \int_0^t \left(\int_0^{t-s} u(t - s - s_1)f'(s_1)ds_1 \right) v(s)ds$.

Interchanging the orders of integration, a legitimate operation because of the absolute convergence of the double integral, we obtain for the second term on the right-hand side of (5.4.7),

$$(5.4.8) \qquad \int_0^t (\int_0^{t-s} u(t-s-s_1)v(s)ds)f'(s_1)ds_1$$

$$= \int_0^t (u(t-s_1)-1)f'(s_1)ds_1 \quad ,$$

using (5.4.2). Combining the results we obtain

$$(5.4.9) \qquad \int_0^t w(t-s)v(s)ds = f(0)\int_0^t u(t-s)v(s)ds$$

$$+ \int_0^t (u(t-s)-1)f'(s)ds$$

$$= f(0)u(t) - f(0) + \int_0^t u(t-s)f'(s)ds$$

$$- \int_0^t f'(s)ds$$

$$= f(0)u(t) - f(t) + \int_0^t u(t-s)f'(s)ds$$

$$= w(t) - f(t) \quad ,$$

which shows that w satisfies (5.4.1).

An integration by parts in (5.4.5) yields

$$(5.4.10) \qquad w(t) = u(0)f(t) + \int_0^t f(t-s)u'(s)ds \quad ,$$

provided that $u'(s)$ exists. Since this formula has a meaning, even if $f(t)$ is not differentiable, it is reasonable to suspect that (5.4.7) yields the solution of (5.4.1) under suitable conditions upon $u(s)$. We shall not discuss this question in further detail here, since it is more properly a part of the theory of renewal equations when Stieltjes, rather than Riemann, integrals are employed. An expression for w of wider validity would be

$$(5.4.11) \qquad w(t) = u(0)f(t) + \int_0^t f(t-s)du(s) \qquad .$$

In many applications of the renewal equation, in mathematical analysis itself, and to problems of physics, engineering, economics and so on, the question of greatest importance is that of the behavior of $u(t)$ as $t \to \infty$. We shall present in what follows some of the techniques which can be used to determine this behavior. Any detailed discussion of the many different situations which can arise is much more involved than might be imagined.

We shall first explore the ramifications of the contour integral representation, then discuss some elementary approaches.

5.5   Use of the Contour Integral Representation

By the use of the Laplace Transform, we can represent the solution of the renewal equation in the form,

$$(5.5.1) \qquad u(t) = \frac{1}{2\pi i} \int_{(b)} \frac{L(f)e^{st}ds}{1 - L(v)} \qquad .$$

We shall consider some simple complex variable techniques to extract more properties of the solution. Suppose initially that the function

$$(5.5.2) \qquad L(f) = \int_0^\infty e^{-st}f(t)dt$$

is a meromorphic function of s, i.e., possessing only poles of finite order in the s-plane and that $L(v)$ is a function with similar properties.

Since $1 - L(v)$ is a meromorphic function of s, by assumption, its singularities will play no role, except perhaps in cancelling those of $L(f)$. The important contributions of the function $1 - L(v)$ will be made by its zeros. Let these be z_1, $z_2, \ldots$ and suppose they can be enumerated in terms of decreasing real part,

(5.5.3) $b > \text{Re}(z_1) > \text{Re}(z_2) > \ldots$

If we shift the contour of integration from the line $b + i\tau$ to the line $b_1 + i\tau$ when $\text{Re}(z_1) > b_1 > \text{Re}(z_2)$, we pick up the residue term of z_1 due to the denominator $1 - L(v)$ and possibly some residues from poles of $L(f)$. Suppose, as is often the case, that $L(f)$ has no singularities in this region. Then (5.5.1) yields,

(5.5.4) $u(t) = k_1 e^{z_1 t} + \int_{(b_1)} \dfrac{L(f) e^{st} ds}{1 - L(v)}$

where k_1 is a constant, given, if z_1 is a simple pole, by

(5.5.5) $k_1 = \dfrac{\displaystyle\int_0^\infty e^{-z_1 t} f(t) dt}{\displaystyle\int_0^\infty t e^{-z_1 t} v(t) dt}$.

Under reasonable conditions, we would suspect that,

(5.5.6) $u(t) = k_1 e^{z_1 t} + O(e^{b_1 t})$.

5.6   Some Important Results

In this section we shall consider the behavior of the solution to the renewal equation as $t \to \infty$.

Consider the equation

(5.6.1) $u(t) = 1 + \displaystyle\int_0^t v(t - s) u(s) ds$

and assume that

(5.6.2)   (a) $v(s) \geq 0$

           (b) $\int_0^\infty v(s)ds < 1$.

Then since $u(t)$ is monotone increasing and bounded, as shown in the previous sections, $u(\infty) = \lim_{t \to \infty} u(t)$ exists. From (5.6.1) we see that

(5.6.3) $u(\infty) = 1 + u(\infty)\int_0^\infty v(s)ds$

           $= \dfrac{1}{1 - \int_0^\infty v(s)ds}$.

Another result can be established immediately. If

(5.6.4)   (a) $v(s) \geq 0$

           (b) $\int_0^\infty v(s)ds = 1,$ $m_1 = \int_0^\infty sv(s)ds < \infty$,

then the solution of

(5.6.5) $u(t) = 1 + \int_0^t u(t-s)v(s)ds$

satisfies

(5.6.6) $u(t) \sim t/m_1$

as $t \to \infty$.

        To establish this fact, in (5.6.5) set $b = 1/m_1$ and $u = bt + w(t)$. Then

$$(5.6.7) \qquad w(t) + bt = 1 + \int_0^t w(t-s)v(s)ds + b\int_0^t (t-s)v(s)ds$$

$$= 1 - b\int_0^t sv(s)ds + bt\int_0^t v(s)ds$$

$$+ \int_0^t w(t-s)v(s)ds \qquad ,$$

or

$$(5.6.8) \qquad w(t) = - bt\left(1 - \int_0^t v(s)ds\right) + 1 - b\int_0^t sv(s)ds$$

$$+ \int_0^t w(t-s)v(s)ds \qquad .$$

Since

$$(5.6.9) \qquad t\left(1 - \int_0^t v(s)ds\right) = t\int_t^\infty v(s)ds \leq \int_t^\infty sv(s)ds \quad ,$$

we see that (5.6.8) may be written in the form

$$(5.6.10) \qquad w(t) = f(t) + \int_0^t w(t-s)v(s)ds \qquad ,$$

where $f(t) \to 0$ as $t \to \infty$, because of assumption (5.6.4).
We now wish to show that $w(t) = O(t)$ as $t \to \infty$. To do this, let
us prove that $|w| \leq a + \varepsilon t$ as $t \to \infty$ where ε is any preassigned
positive constant and $a = a(\varepsilon)$. Consider the solution of (5.6.10),
as obtained by the method of successive approximations

$$(5.6.11) \qquad w_0 = f \; ,$$

$$w_{n+1} = f + \int_0^t w_n(t-s)v(s)ds \qquad .$$

Let us now choose t_0 with the condition that $|f| \leq \varepsilon$

for $t \geq t_0$ and $t_0 \geq 1$. Let $a_0 = \max |f|$ in $[0, t_0]$ if this maximum is nonzero, otherwise, it is equal to 1. Then clearly for all $t \geq 0$, we have $|w_0| \leq a_0 + \varepsilon t$. Using this bound in w_1 as given by (5.6.11), we obtain in $[(0, t_0)]$

$$(5.6.12) \quad |w_1| \leq a_0 + \int_0^t (a_0 + \varepsilon(t-s))v(s)ds$$

$$\leq a_0 + a_0 \int_0^{t_0} v(s)ds + \varepsilon t_0 \int_0^{t_0} v(s)ds$$

$$- \varepsilon \int_0^t sv(s)ds$$

$$\leq a_0 + a_0 \int_0^{t_0} v(s)ds + \varepsilon t_0 \quad ,$$

since

$$\int_0^\infty v(s)ds = 1 \quad \text{and} \quad v(s) \geq 0 .$$

For $t \geq t_0$, we obtain

$$(5.6.13) \quad |w_1| \leq \varepsilon + \int_0^t (a_0 + \varepsilon(t-s))v(s)ds$$

$$\leq \varepsilon + a_0 \int_0^t v(s)ds + \varepsilon t \int_0^t v(s)ds$$

$$\leq \varepsilon + a_0 + \varepsilon t .$$

Let us define

$$a_1 = a_0 + a_0 \int_0^{t_0} v(s)ds .$$

If ε is small enough and t_0 is large enough, we have $a_1 > a_0 + \varepsilon$. We see then that $|w_1| \leq a_1 + \varepsilon t$, for $t \geq 0$.

All the requirements for an inductive proof are now at hand. If we have $|w_n| \leq a_n + \varepsilon t$ for $t \geq 0$, the same argument as above yields $|w_{n+1}| \leq a_{n+1} + \varepsilon t$, where

$$(5.6.14) \qquad a_{n+1} = a_0 + a_n \int_0^{t_0} v(s)ds \qquad .$$

If v is not identically zero for $t \geq t_0$, the conditions

$$v \geq 0 \qquad \text{and} \qquad \int_0^\infty v(s)ds = 1$$

yield

$$\int_0^{t_0} v(s)ds < 1 \qquad ,$$

and hence,

$$(5.6.15) \qquad a_n < a_\infty = \frac{a_0}{1 - \int_0^{t_0} v(s)ds} \qquad .$$

if v is identically zero for $t \geq t_1$, for some t_1, there is no difficulty in obtaining the asymptotic behavior of u by other means, since

$$(5.6.16) \qquad 1 - \int_0^\infty v(s)e^{-st}dt = 1 - \int_0^{t_1} v(s)e^{-st}dt$$

is now an entire function. Hence, we may, with impunity assume that

$$\int_0^{t_0} v(s)ds < 1$$

for any fixed finite t_0.

Since $|w_n| < a_n + \varepsilon t < a_\infty + \varepsilon t$ for all n and for $t \geq 0$, it follows that the solution enjoys the same property, which means that $w(t) = 0(t)$ as $t \to \infty$ since ε is arbitrary.

5.7 Systems of Renewal Equations

In this section we wish to study the solution of systems of linear integral equations of the form

$$(5.7.1) \qquad u_i(t) = f_i(t) + \int_0^t \sum_{j=1}^N u_j(t-s)b_{ij}(s)ds \quad,$$

$$i = 1,2,\ldots,N \quad.$$

Although the result concerning existence and uniqueness of solutions extend in a routine fashion, new and powerful methods are required to handle the questions of asymptotic behavior of the solutions as $t \to \infty$. In most investigations of equations of this nature, in analysis and in mathematical physics alike, this is the property of the solution of most significance.

Since any detailed investigation requires complicated and sustained analysis, we shall content ourselves here with a discussion of a typical result that can be obtained and a sketch of the general method that can be employed.

Letting $u(t)$ represents a column vector whose components are $u_i(t)$ $(i = 1,2,\ldots,N)$ and $B(t) = (b_{ij}(t))$, we can write (5.7.1) in the form,

$$(5.7.2) \qquad u(t) = f(t) + \int_0^t B(s)u(t-s)ds \quad.$$

Proceeding formally, use of the Laplace Transform yields the equation

$$(5.7.3) \qquad L(u) = L(f) + L(B)L(u) \quad,$$

or

$$(5.7.4) \qquad L(u) = (I - L(B))^{-1}L(f) \qquad .$$

We see then that the asymptotic behavior of the vector $u(t)$ as $t \to \infty$ will depend upon the location and multiplicity of the zeros of the determinental equation

$$(5.7.5) \qquad \left| I - \int_0^\infty e^{-st}B(t)dt \right| = 0 \qquad .$$

Any direct investigation of this equation appears to be very difficult. This problem can be attacked by using the theory of positive matrices.

5.8   Branching Processes

The renewal equation occurs in many fields of mathematical analysis and, in particular, we wish to show in this section that it occurs quite naturally in the field of branching processes.

Consider a branching process which can be described in the following simple way. Consider an object born at time $t = 0$ with a random life length given by the probability distribution function $G(t)$. At the end of its life, it is replaced by a random number of similar objects, with p_r being the probability that the new number of objects is r, $r = 1, 2, \ldots$ Their probabilities are taken to be independent of absolute time, the age of the object when it dies and of the number of objects present at time t.

The process is begun at time $t = 0$ with a single object or particle which initiates a cascading or branching process as time advances and we are interested in deriving an equation governing the expected number of particles resulting in the process.

Using the assumptions stated above, at any time t the expected number of particles created is given by,

$$(5.8.1) \qquad E(n) = \sum_{n=1}^{\infty} n p_n \qquad .$$

We can now define a generating function associated with this expectation by,

$$(5.8.2) \qquad h(s) = \sum_{n=1}^{\infty} p_n s^n, \qquad |s| \leq 1$$

which is a converging series within the unit circle in the complex plane. It is immediately clear that

$$(5.8.3) \qquad \frac{dh(s)}{ds}\bigg|_{s=1} = \sum_{n=1}^{\infty} n p_n \qquad .$$

We also know from our assumptions that $G(t)\Delta t$ is given as the probability that a particle dies within the interval $(t, t+\Delta t)$.

Let $z(t)$ be the number of particles present at time t. This is a random function which we call an age-dependent branching process.

Let

$$(5.8.4) \qquad f_n(t) = \text{Prob}(z(t) = n)$$

i.e., we define $f_n(t)$ to be the probability that the number of existing particles is n at time t.

Again we can define a generating function in the complex plane as,

$$(5.8.5) \qquad f(s, t) = \sum_{n=1}^{\infty} f_n(t) s^n$$

and we can write down a nonlinear integral equation

$$(5.8.6) \qquad f(s, t) = s(1 - G(t)) + \int_0^t h(f(s, t - t_1)) dG(t_1)$$

by arguing that the expected number of particles, represented by the generating function $f(s, t)$ must equal those particles not destroyed, $s(1 - G(t))$ plus those created over the time interval $(0, t)$.

Taking the derivative of (5.8.6) with respect to s and evaluating the result as $s = 1$, we have

$$(5.8.7) \qquad E(z(t)) = (1 - G(t)) + h'(1)\int_0^t E(z(t - t_1))dG(t_1) \quad .$$

Finally, if we can write $dG(t) = p(t)dt$, we have

$$(5.8.8) \qquad E(z(t)) = (1 - G(t)) + h'(1)\int_0^t E(z(t - t_1))p(t_1)dt_1$$

which is the renewal equation.

Problems

1. Consider the equation

$$\partial u/\partial t + ixu - iaf(x)\int_0^\infty u\,dx = 0, \qquad u(x, 0) = h(x) \quad .$$

Let

$$v(t) = \int_{-\infty}^\infty u(x, t)dx \quad ,$$

$$H(t) = \int_{-\infty}^\infty e^{-ixt}h(x)dx \quad ,$$

$$K(t) = ia\int_{-\infty}^\infty f(x)e^{-ixt}dx \quad .$$

Show that

$$v(t) = H(t) + \int_0^t K(t - s)ds \quad .$$

(Bellman, R., and Richardson, J.M., "On the Stability of the Linearized Plasma Theory", J. Math. Analysis and Appl., 1, 308-313 (1960).

2. What is the solution of $u(t) = f(t) + \int_0^t u(t-s)v(s)ds$ when $v(s)$ is an exponential polynomial of the form

$$v(s) = \sum_{k=1}^N p_k(s)e^{\lambda_k s} \quad,$$

where the $p_k(s)$ are polynomials in s?

3. Solve the equation

$$md^2u/dt^2 + a^2u(t) + \int_0^t K(t-s)du(s)/dsds = q(t) \quad .$$

4. Solve

$$u(x) + c_1\int_0^x \exp(a_1(x-y))u(y)dy = b(x)$$

and generally

$$u(x) + \int_0^x \sum_{k=1}^N c_k\exp(a_k(x-y)))u(y)dy = b(x) \quad .$$

5. Solve the equation

$$u(t) + \int_0^t su(t-s)ds + \int_0^1 u(s)ds = 1 \quad, \qquad t \geq 0 \quad .$$

6. Let $f(t)$ denote the proportion of an original quantity of goods remaining unsold at time t after their purchase. Let $v(t)$ denote the rate of purchase of goods to replenish the stock. Assume that the original stock is 1 unit and that $v(t)$ is to be adjusted to maintain a constant stock. Show that

$$1 = f(t) + \int_0^t f(t-x)v(x)dx \quad .$$

Assuming that goods are sold at a constant rate, so that

$$f(t) = \begin{cases} 1 - t/T & t < T \\ 0 & t > T \end{cases} \quad ,$$

show that the Laplace Transform of v is given by

$$(1 - e^{-sT})(Ts - 1 + e^{-sT})^{-1} \quad .$$

By expanding the second factor in inverse powers of $(Ts - 1)e^{st}$, deduce the solution $v(t)$ in each interval $nT < t < (n+1)T$, $n = 0,1,\ldots$

(Bateman, H., "An Integral Equation Occurring in a Mathematical Theory of Retail Trade", *Messenger of Math.*, <u>49</u>, 134-137 (1920).

7. Use the Laplace Transform to solve the matrix equation

$$dX/dt = X(+0)X(t) - \int_0^t K(s)X(t-s)ds$$

and to show that $X(t)$ also satisfies the equation

$$dX/dt = X(t)X(+0) - \int_0^t X(t-s)K(s)ds \quad .$$

Chapter 6

NUMERICAL INVERSION OF THE LAPLACE TRANSFORM

6.1 Introduction

In the previous chapters we have defined and used the analytic definition of the Laplace Transform in several important applications. We have seen that if $f(t)$ is defined for $t \geq 0$, then the Laplace Transform is defined as,

$$(6.1.1) \qquad F(s) = \int_0^\infty f(t)e^{-st}dt$$

and the inverse transform is

$$(6.1.2) \qquad f(t) = \frac{1}{2\pi i} \int_{-i\infty+c}^{i\infty+c} F(s)e^{st}ds$$

where the contour is a path taken in the complex plane, $s = \tau + i\tau$.

We have considered one of the successful applications of the Laplace Transform in which a linear ordinary differential equation with constant coefficients can be reduced to an algebraic equation defining the transform function $F(s)$ in the complex plane. There remains only the problem of determing $f(t)$ knowing the complex function $F(s)$.

This well posed problem offers many interesting challenges to the mathematician in finding analytical and numerical techniques of solutions. In examining this problem, we are faced with several possible approaches to a solution. We could use (6.1.2) as a direct definition of the unknown function f(t) and compute the analytical solution by evaluation of the contour integral. This approach has been examined in previous chapters and relies on the calculus of residues. Extensive tables of the Laplace Transforms and their inverses exist to aid this approach.

We can also regard (6.1.1) as an integral equation for the function f(t) where the transform function F(s) is known throughout the complex plane. This approach, as we shall show, gives rise to many fruitful and interesting numerical results.

The basic idea which we shall explore for the numerical inversion of the Laplace Transform is to use the method of numerical quadrature to reduce the integral Eq. (6.1.1) to a system of linear algebraic equations. A detailed examination of the techniques will reveal several numerical problems which, as we shall see, will require some sophisticated ideas.

6.2   The Complex Laplace Transform

We begin by recalling the Laplace Transform

$$(6.2.1) \qquad F(s) = \int_0^\infty e^{-st} f(t) dt$$

where s is an arbitrarily chosen complex parameter. This implies that if s is chosen to be complex, then the function F(s) will be complex. The first thing which strikes one viewing (6.2.1) as an integral equation, is the limits of integration from zero to infinity. From a numerical standpoint this is quite burdensome and by the transformation $r = e^{-t}$ (6.2.1) can be transformed to

$$(6.2.2) \qquad F(s) = \int_0^1 r^{s-1} g(r) dt$$

where $g(r) = f(-\log(t))$.

Keep in mind the transformation of points in the t and s ranges in (6.2.1). Points in the neighborhood of r = 0 reflect the behavior of f(t) for large time. Those in the neighborhood of r = 1 gives the behavior of the function f(t) near t = 0.

The solution of the integral equation numerically depends on the choice of s in the complex plane and we are immediately faced with the problem of selecting s. A clue to the numerical process is found by considering the form of (6.2.2) as a complex function. Since r is real and s is a complex parameter, then it can easily be shown that

$$(6.2.3) \qquad r^{s-1} = r^{((\sigma-1) + i\tau)}$$

$$= r^{(\sigma-1)} \, r^{i\tau}$$

$$= r^{(\sigma-1)} \, e^{i\tau \log r}$$

If F(s) is resolved into its real and imaginary parts,

$$(6.2.4) \qquad F(s) = a(\sigma,\tau) + ib(\sigma,\tau)$$

$$a(\sigma,\tau) = \int_0^t r^{(\sigma-1)} \cos(\tau \log r)g(r)dt$$

$$b(\sigma,\tau) = \int_0^1 r^{(\sigma-1)} \sin(\tau \log r)g(r)dt \quad .$$

Numerically, Eqs. (6.2.4) are very interesting. If we insist on the freedom of choosing s in the complex plane, we must solve two real linear integral equations. Furthermore, if $\sigma < 1$, the integrand has a singularity at r = 0, making numerical computation difficult if not impossible. In addition if $\tau \neq 0$, the integrand fluctuates with increasing rapidity near the origin and is, in fact, unbounded when $\sigma < 1$.

To avoid these problems, we shall choose s to lie on the positive real axis greater than unity. In fact we shall later select s to be the positive real integers, 1,2,3,...

6.3 Numerical Quadrature

The numerical techniques which we shall explore in this section are motivated by our problem of solving the transformed integral equation

$$(6.3.1) \qquad \int_0^1 r^{s-1} g(r)dr = F(s) \quad .$$

To this end, let us assume that we are able to deal with approximations of the general form,

$$(6.3.2) \qquad \int_a^b f(x)ds = \sum_{i=1}^{N} w_i f(x_i)$$

where both the weights w_i and the values of x, x_i, must be determined for the most accurate approximation.

If the limits of integration in (6.3.2) are finite, a linear transformation will reduce (6.3.2) to the form

$$(6.3.3) \qquad \int_{-1}^1 f(x)dx = \sum_{i=1}^{N} w_i f(x_i) \quad .$$

In Eq. (6.3.3), two adjustable parameters w_i and x_i are introduced in the approximation. Since (6.3.3) is the approximate relation, the first problem is to determine the parameters w_i and x_i so as to obtain a reasonable approximation. To do this we shall follow a procedure inaugurated by Gauss. We require that (6.3.3) be exact for any polynomial of degree less than or equal to (2N-1).

Let us assume that there exists a set of polynomials $P_N(x)$ of order N $(-1 \le x \le 1)$ satisfying the following $(N+1)$ conditions:

$$(6.3.4) \qquad \int_{-1}^1 r^m P_N(r)dr = 0 , \qquad m = 0,1,\ldots,N-1 \quad ,$$

$$(6.3.5) \qquad \int_{-1}^{1} r^N P_N(r)dr = 1 \qquad .$$

In the next section we shall show that these polynomials exist, are unique, and have several important properties.

If we assume for the moment that these polynomials do exist, then if we set $f(r) = r^m P_N(r)$, for a given N and $m = 0,1,\ldots,N-1$, then the Gauss condition that the approximate relation (6.3.3) be exact will allow us to determine both the set of points r_i and the associated weights w_i.

For the trial functions $f(r) = r^m P_N(r)$, substituting into (6.3.3) and using (6.3.4) we obtain a set of N linear algebraic equations,

$$(6.3.6) \qquad 0 = \int_{-1}^{1} r^m P_N(r)dr = \sum_{i=1}^{N} w_i r_i^k P_N(r_i)$$

$$k = 0,1,\ldots,N-1 \qquad .$$

The test polynomials $r^k P_N(r), \, k = 0,1,\ldots,N-1$ are all of degree $2N-1$ or less. If the system (6.3.6) is considered to be a set of equations for the N quantities $w_i f(x_i)$, it can be written in matrix form as,

$$(6.3.7) \qquad
\begin{bmatrix}
1 & 1 & \cdots & 1 \\
r_1 & r_2 & \cdots & r_N \\
r_1^2 & r_2^2 & \cdots & r_N^2 \\
r_1^{N-1} & r_2^{N-1} & \cdots & r_N^{N-1}
\end{bmatrix}
\cdot
\begin{bmatrix}
w_1 P_N(r_1) \\
w_2 P_N(r_2) \\
\cdot \\
w_N P_N(r_N)
\end{bmatrix}
= 0 \qquad .$$

The matrix in (6.3.7) is the Vandermonde matrix whose determinant can be shown to be,

$$(6.3.8) \quad \begin{vmatrix} 1 & 1 & . & . & . & 1 \\ r_1 & r_2 & . & . & . & r_N \\ r_1^2 & r_2^2 & . & . & . & r_N^2 \\ & & & & & \\ r_1^{N-1} & r_2^{N-1} & . & . & . & r_N^{N-1} \end{vmatrix} = \prod_{(1 \le i \le j < N)} (r_j - r_i) \ .$$

If $r_i \ne r_j$, the determinant is nonzero and the only solution to (6.3.7) is

$$(6.3.9) \quad w_i P_N(r_i) = 0, \qquad i = 1,2,\dots,N \quad .$$

Since the weighting functions are assumed to be nonzero, the only solution to (6.3.7) is

$$(6.3.10) \quad P_N(r_i) = 0, \qquad i = 1,2,\dots,N \quad .$$

Therefore the set r_i is the set of zeros of the polynomial, $P_N(r)$.

To determine the weighting functions, w_i, we shall use a different test polynomial, namely

$$(6.3.11) \quad f(r) = \frac{P_N(r)}{(r-r_i)P_N{}'(r_i)}$$

then

$$(6.3.12) \quad f(r_j) = 0 \ ,$$
$$f(r_i) = 1 \ , \qquad j \ne i \quad .$$

Substituting (6.3.11) into (6.3.2) gives,

$$(6.3.13) \quad w_i = \int_{-1}^{1} \frac{P_N(r)}{(r-r_i)P_N{}'(r_i)} \, dr \quad .$$

This establishes the necessity of the condition of (6.3.10) and (6.3.13). Let us demonstrate the sufficiency.

We wish to show that

$$(6.3.14) \qquad \int_{-1}^{1} g(r)dr = \sum_{i=1}^{N} w_i g(r_i)$$

if $g(r)$ is a polynomial of degree $2N-1$ and w_i and r_i are given above. Let

$$(6.3.15) \qquad g(r) = g_1(r)P_N(r) + g_2(r)$$

where $g_1(r)$ and $g_2(r)$ are both polynomials of degree of at most $N-1$.

Then we must show that

$$(6.3.16) \qquad \int_{-1}^{1} g_1(r)P_N(r)dr + \int_{-1}^{1} g_2(r)dr$$

$$= \sum_{i=1}^{N} w_i g_1(r_i)P_N(r_i) + \sum_{i=1}^{N} w_i g_2(r_i)$$

but by (6.3.4) and the choice of r_i, (6.3.16) reduces to,

$$(6.3.17) \qquad \int_{-1}^{1} g_2(r)dr = \sum_{i=1}^{N} w_i g_2(r_i)$$

which must be shown to be true for a polynomial of degree of at most $N-1$. Using the definition of w_i,

$$(6.3.18) \qquad \sum_{i=1}^{N} w_i g_2(r_i) = \sum_{i=1}^{N} \int_{-1}^{1} \frac{P_N(r)g_2(r_i)}{(r - r_i)P_N'(r_i)} dr$$

$$= \int_{-1}^{1} \sum_{i=1}^{N} \frac{P_N(r)g_2(r_i)dr}{(r - r_i)P_N'(r_i)} \quad .$$

Consider the polynomial

$$(6.3.19) \qquad \sum_{i=1}^{N} \frac{P_N(r) g_2(r_i)}{(r - r_i) P_N'(r_i)} \qquad .$$

It takes on the values of $g_2(r_i)$ at the points $r = r_i$ and by the Lagrange interpolation formula for $g_2(r)$, $g_2(r)$ is of degree N-1. If $g_2(r)$ is of degree less than N-1, then consider a new polynomial $g_2(r) + r^{N+1}$ and use the linearity of both sides of (6.3.14).

By the foregoing argument, we have established the fact that if $g_2(r)$ is a polynomial of degree less than or equal to 2N-1, its integral can be exactly represented by

$$\int_{-1}^{1} g(r)dr = \sum_{i=1}^{N} w_i g(r_i)$$

where

$$w_i = \int_{-1}^{1} \frac{P_N(r)dr}{(r - r_i) P_N'(r_i)}$$

and the parameters r_i are defined by the condition,

$$P_N(r_i) = 0 , \qquad i = 1, 2, \ldots, N \qquad .$$

$P_N(r)$ is a polynomial of degree N satisfying conditions (6.3.4) and (6.3.5). Such an interpretation of the integral (6.3.17) is known as a numerical quadrature and will be used in our formulation of the numerical inversion of the Laplace Transform.

In the next section we shall demonstrate that the polynomials $P_N(r)$ do, indeed, exist and from their properties we will show them to be the Legendre polynomials.

6.4   The Legendre Polynomials

Let $P_N(r)$ be an N-th order polynomial satisfying the following N+1 conditions,

$$(6.4.1) \qquad \int_{-1}^{1} P_N(r) r^m dr = 0 \ , \qquad m = 1, 2, \ldots, N-1$$

$$\int_{-1}^{1} P_N(r) r^N dr = 1 \quad .$$

We begin our study of these polynomials by observing that if $P_N(r)$ exists, then it is unique. Let $P_n(r)$ and $q_n(r)$ both satisfy conditions (6.4.1).

Then because of linearity,

$$(6.4.2) \qquad \int_{-1}^{1} g(r)(p_n(r) - q_n(r))dr = 0$$

for any polynomial $g(r)$ of degree less than or equal to n. Let $g(r) = (p_n(r) - q_n(r))$. Then

$$(6.4.3) \qquad \int_{-1}^{1} ((p_n) - q_n(r))^2 dr = 0 \quad ,$$

which can only be true if $p_n(r) = q_n(r)$.

To show that the polynomials exist, we construct them in such a way that the conditions stated in (6.4.1) are satisfied.

Consider the polynomial of degree N defined in the following way,

$$(6.4.4) \qquad \pi_N(r) = \left(\frac{d}{dr}\right)^N (r^2 - 1)^N \quad .$$

By repeated integrations by parts, it can be shown that

$$(6.4.5) \qquad \int_{-1}^{1} r^m \pi_N(r)dr = 0 \ , \qquad m = 0, 1, \ldots, N-1$$

and

$$(6.4.6) \qquad \int_{-1}^{1} r^N \pi_N(r)dr = (-1)^N N! \int_{-1}^{1} (1 - r^2)^N dr \neq 0 \quad .$$

144

For convenience let

(6.4.7) $\quad P_N(r) = \dfrac{1}{2^N N!}\left(\dfrac{d}{dr}\right)^N (r^2-1)^N$.

In fact after repeated differentiations indicated by (6.4.7)

(6.4.8) $\quad P_N(r) = \displaystyle\sum_{k=0}^{m} (-1)^k \dfrac{(2N-2k)!}{2^N k!(N-k)!(N-2k)!} r^{(N-k)}$

where $m = N/2$ or $(N-1)/2$, whichever is an integer.

Integration by parts yields the result,

(6.4.9) $\quad \displaystyle\int_{-1}^{1} P_N^2(r)dr = \dfrac{2}{2N+1}$.

From a computational standpoint (6.4.9) becomes hard to evaluate for large N because of the oscillations of the terms. We can avoid such dangerous numerical procedures by considering the use of recursive relationships.

Using (6.4.7), it is not difficult to show that the following two recursive relations hold,

(6.4.10) $\quad (N+1)P_{N+1}(r) - (2N+1)rP_N(r) + NP_{N-1}(r) = 0$,

$\qquad\qquad\qquad\qquad\qquad\qquad N = 1,2,\ldots$

and

(6.4.11) $\quad (r^2-1)P_N'(r) = NrP_N(r) - NP_{N-1}(r)$, $\quad N = 1,2,\ldots$

Furthermore, $P_N(r)$ satisfies the second-order linear differential equation,

(6.4.12) $\quad (1-r^2)\dfrac{d^2u}{dr^2} - 2r\dfrac{du}{dr} + N(N+1)u = 0$.

The condition (6.4.5) implies the very important property of orthogonality over their interval $(-1, 1)$, i.e.,

(6.4.13) $\displaystyle\int_{-1}^{1} P_m(r)P_n(r)dr = 0$, $m \neq n$

$$\int_{-1}^{1} P_n^2(r)dr = \frac{2}{2n+1} \quad .$$

The aforementioned properties establish $P_N(r)$ to be the well-known Legendre polynomials.

We shall now show that the zeros of $P_N(r)$ are real, distinct and are contained in the interval $(-1, 1)$. These properties are crucial to the success of the numerical quadrature technique discussed in the last section.

We argue from the direct definition of $P_N(r)$ given in (6.4.1). The polynomial $P_N(r)$ must change sign at least once in the interval $(-1, 1)$ for we know,

$$\int_{-1}^{1} P_N(r)dr = 0 \quad .$$

Let $r_1, r_2, \ldots, r_m$, $1 < m < N$ be the roots of odd order in $(-1, 1)$. Suppose $m < N$. Then the expression

$$\int_{-1}^{1} (r - r_1), (r - r_2), \ldots, (r - r_m)P_N(r)dr$$

must be nonzero. But this contradicts the condition $m < N$ in (6.4.1).

6.5   Numerical Inversion of the Laplace Transform

We are now in a position to develop an algorithm for the numerical inversion of the Laplace Transform. Our aim is to determine the values of $u(t)$ as accurately as possible from the values of $F(s)$, $s > 0$, known to a prescribed degree of accuracy. As before, $F(s)$ is the Laplace Transform of $u(t)$.

$$(6.5.1) \qquad F(s) = \int_0^\infty e^{-st} u(t) dt \qquad .$$

The basic assumptions we make are twofold: first, that this is a well-posed problem in the sense that an exact determination of $F(s)$ would lead to an exact, hence unique, determination of $u(t)$; second, that $u(t)$ is sufficiently smooth to permit the approximation methods we employ. By making the change of variable $r = e^{-t}$, we can transform (6.5.1) into a linear integral equation over a finite range,

$$(6.5.2) \qquad \int_0^1 r^{s-1} u(-\log r) \, dr = F(s) \qquad .$$

Setting $g(r) = u(-\log r)$ we have the integral equation,

$$(6.5.3) \qquad \int_0^1 r^{s-1} g(r) dr = F(s)$$

to which we wish to apply the quadrature formula derived in the last section. Since the quadrature formula we need is over the interval $[0, 1]$, we shall replace the Legendre polynomials designed for the interval $[-1, 1]$ by the shifted Legendre polynomials defined by,

$$(6.5.4) \qquad p^*_N(r) = p_N(1 - 2r) \qquad .$$

Therefore, we can write

$$(6.5.5) \qquad \sum_{i=1}^N w_i r_i^{s-1} g(r_i) = F(s) \qquad .$$

Letting s assume N different values, say $s = 1, 2, \ldots, N$, yields a linear system of N equations in N unknowns, $g(r_i)$, $i = 1, 2, \ldots, N$. Therefore

$$(6.5.6) \qquad \sum_{i=1}^N w_i r_i^k g(r_i) = F(k+1) , \qquad k = 0, 1, \ldots, N \qquad ,$$

where r_i are the roots of the shifted Legendre polynomial, $P^*_N(r) = 0$ and

$$(6.5.7) \qquad w_i = \int_0^1 \frac{P^*_N(r)dr}{(r - r_i)P^{*'}_N(r_i)} \quad .$$

These N equations uniquely determine N quantities $w_i g(r_i)$ since the determinant of the coefficients is the Vandermonde determinant which is nonzero.

6.6   Explicit Inversion Formula

Since the success of the numerical inversion process depends on the inversion of a rather large Vandermonde matrix, it is our advantage to seek an analytic expression for the inverse. Let us introduce a new variable, $y_i = w_i g(r_i)$, $i = 1,2,\ldots,N$ and set $a_k = f(k + 1)$.

Then (6.5.6) takes the form

$$(6.6.1) \qquad \sum_{i=1}^{N} r_i^k y_i = a_k , \qquad k = 0,1,\ldots,N \quad .$$

Using classical methods, we multiply the k-th equation by a parameter, q_k, to be determined, and sum the results over k. Therefore,

$$(6.6.2) \qquad \sum_{i=1}^{N} \left(y_i \sum_{k=0}^{N-1} q_k r_i^k \right) = \sum_{k=0}^{N-1} q_k a_k \quad .$$

Let the polynomial $f(r)$ be defined as

$$(6.6.3) \qquad f(r) = \sum_{k=0}^{N-1} q_k r^k \quad .$$

Thus we have

$$(6.6.4) \qquad \sum_{i=1}^{N} y_i f(r_i) = \sum_{k=0}^{N-1} q_k a_k \quad .$$

We are at liberty to choose $f(r)$ in a convenient manner. To this end, suppose we demand that $f(r) = f_j(r)$ be chosen so that the orthogonality conditions

$$(6.6.5) \qquad f_j(r_i) = 0, \qquad i \neq j$$

$$f_j(r_j) = 1$$

are true.

Since $f_j(r)$ is a polynomial, it can be written as,

$$(6.6.6) \qquad f_j(r) = \sum_{k=0}^{N-1} q_{kj} r^k \quad .$$

Substituting into (6.6.4) yields

$$(6.6.7) \qquad y_j = \sum_{k=0}^{N-1} q_{kj} a_k \quad ,$$

where q_{kj}, $k = 0, 1, \ldots, N-1$ are determined from condition (6.6.5).

Because of the fact that the r_i are the zeros of the shifted Legendre polynomial, the polynomial satisfying (6.6.5) is determined by the Lagrange interpolation formula,

$$(6.6.8) \qquad f_j(r) = \frac{p^*_N(r)}{(r - r_j) p^{*\prime}_N(r_j)} \quad .$$

The desired q_{kj} in (6.6.7) are the coefficients of this polynomial of degree $N-1$. Repeating this procedure for each j yields the coefficients of the inverse Vandermonde matrix.

With this information, we can compute the values of $y_i = w_i g(r_i)$ at the zeros of the shifted Legendre polynomials. After computing the values of the weighting factors, w, and shifting back to the time axis, the values of $u(t_k)$ can be determined.

6.7   Instability of the Inverse of the Laplace Transform

For several reasons we cannot expect that any specific method for the inversion of the Laplace Transform will work equally well in all cases. The mathematical reason for this is that the Laplace inverse is an unbounded operator. In other words, arbitrary small changes in $F(s)$ will produce arbitrary large changes in the values of u.

This instability manifests itself in the behavior of the matrix $(w_i r_i^k)$, $i = 1,2,\ldots,N$, $k = 0,1,\ldots,N-1$, which is ill-conditioned. We shall use the term ill-conditioned to describe a non-singular matrix A when A^{-1} contains elements of both signs of large magnitude in such a way that a small change in the vector b can produce a large change in the solution of $Ax = b$. In other words, x is an unstable function of b. The ill-conditioning of $(w_i r_i^k)$ rapidly worsens as N increases.

To illustrate the point, consider the transform of $\sin at$,

$$(6.7.1) \qquad \int_0^\infty e^{-st} \sin at\, dt = \frac{a}{a^2 + s^2} \quad .$$

The Laplace Transform is uniformly bounded by $1/a$ for $s > 0$. However, the function $\sin at$ oscillates between $|+1, -1|$ with increasing rapidity as a gets large.

When we decide to perform a numerical inversion of the Laplace Transform, we are assuming that the function $u(t)$ is actually quite smooth. Our fundamental requirement is that $u(-\log r)$ can be well approximated by a polynomial in r, $0 \le r \le 1$.

At this time our concept of "well approximated" depends on a mean square approximation. There is no reason why a more sophisticated method cannot be applied, nor is the polynomial approximation the only means of representing $u(-\log r)$. Nevertheless, we are faced with the problem of computing the inverse of an ill-conditioned matrix. In the next section we shall explore techniques to accomplish this.

6.8   Tychonov Regularization

In this section our problem is to devise a computational algorithm, for obtaining an acceptable solution to

(6.8.1)   Ax = b

where the inverse of the $N \times N$ system A^{-1}, exists but is ill-conditioned. The first idea is to regard (6.8.1) as the result of minimizing the quadratic form

(6.8.2) $Q(x) = (Ax - b, Ax - b)$

where (,) is the vector inner product and thus replace (6.8.1) by

(6.8.3) $A^T Ax = A^T b$.

In this formulation, the matrix of coefficients is symmetric.

In this form the matrix $A^T A$ may be more ill-conditioned than A. However, suppose we modify (6.8.2) by replacing the solution of (6.8.1) by the problem of minimizing.

(6.8.4) $R(x) = (Ax - b, Ax - b) + h(x)$

where $h(x)$ is a function chosen to insure the stability of the equation for minimizing over x. This is called Tychonov regularization.

The rationale behind this is the following. We accept the fact that the general problem of solving (6.8.1), when A is ill-conditioned and b is imprecisely given, is impossible. We cannot guarantee accuracy. In reality (6.8.1) determines a set of vectors for which $\|Ax - b\| \leq \varepsilon$ where $\|...\|$ is some suitable norm. Additional information is required to select an acceptable solution. The function $h(x)$ represents the additional information but the question still remains as to how $h(x)$ is to be defined.

Let us give an example which we will expand upon. Suppose,

as a result of preliminary information, we know $x - c$, a given vector. Let

$$(6.8.5) \qquad h(x) = \lambda(x - c, x - c)$$

where $\lambda > 0$ and consider the problem of minimizing

$$(6.8.6) \qquad R(x, \lambda) = (Ax - b, Ax - b) + \lambda(x - c, x - c) \qquad .$$

It is easy to see that the minimum value of x is given by

$$(6.8.7) \qquad x(\lambda) = (A^T A + \lambda I)^{-1}(A^T b + \lambda c) \qquad .$$

We can observe immediately that if λ is small, we are close to the desired value $A^{-1}b$ but $(A^T A + \lambda I)$ is still ill-conditioned. As λ gets large the ill-conditioning of the matrix is relieved, but the solution is not accurate. The choice of λ is by trial and error, but by (6.8.1) we can always check the goodness of fit.

If we consider the vector c to be the first approximation to the solution x, we can develop an iteration scheme by setting

$$(6.8.8) \qquad x_0 = c$$

$$x_{n+1} = (A^T A + \lambda I)^{-1}(A^T b + \lambda x_n) \qquad .$$

It can be shown that x_n will converge for any $\lambda > 0$ and any c.

6.9   Obtaining the Initial Approximation

One way to obtain an initial approximation, as far as the numerical inversion of the Laplace Transform is concerned, is to begin with a low order quadrature approximation. Suppose, for example, we use initially a five-point Gaussian quadrature. Using the Lagrange interpolation formula, we pass a polynomial of degree four through these points $(r_i, g(r_i))$ and then use this polynomial to obtain starting values for the values of g at the seven quadrature points for a quadrature formula of degree seven. The seven values constitute the initial vector c in the successive approximation scheme discussed in the last section. We can continue in this way, increasing the dimension of the process step-by-step until we obtain sufficient agreement.

An alternative procedure would be to set the initial vector $c = 1$ and select $\lambda^{(1)}$ large enough to reduce the ill-conditioning of the matrix $(A^T A + \lambda I)$. By the successive approximation scheme, obtain a convergent vector $x_n^{(1)}$. By setting $c = x_n^{(1)}$ and selecting $\lambda^{(2)} < \lambda^{(1)}$, we can repeat the process to produce $x_n^{(2)}$. At each stage of the process we can compute a "goodness of fit" by calculating

$$(6.9.1) \qquad h(\lambda) = (Ax_n^{(p)} - b)^T (Ax_n^{(p)} - b) \quad .$$

6.10   Change of the Time Scale

The procedure described for the numerical inversion of the Laplace Transform provides values of $u(t)$ at the times $t_k = -\log r_k$ where r_k are the roots of the shifted Legendre polynomial $p_N^*(r)$. It is not difficult to show that the roots of $p_N^*(r)$ are uniformly distributed over $|0, 1|$ to a higher and higher degree as N approaches infinity. Unfortunately, the logarithm of $1/r_i$ does not possess the same distribution property over $[0, \infty]$. The t_k values tend to bunch around $t = 0$ and furnish meagre information for large t.

Hence, we face a problem if we wish to determine $u(t)$ over a more extensive range. We can occasionally use the contour integral to obtain a useful analytic expression for $u(t)$ for large t. In general, however, we must use the numerical techniques.

Let us discuss some rudimentary techniques. First we observe that since

$$(6.10.1) \qquad L(u(t)) = \int_0^\infty e^{-st} u(t) dt = F(s) \quad ,$$

then

$$(6.10.2) \qquad L(u(at)) = \int_0^\infty e^{-st} \, u(at)dt = \frac{1}{a}\int_0^\infty e^{-st/a} \, u(t)dt$$

$$= \frac{1}{a} F(s/a) \qquad .$$

Hence, if we use, in place of $F(1)$, $F(2),\ldots,$ the values of $F(1/a)$, $F(2/a),\ldots,$ we can determine, via the approximation formula,

$$(6.10.3) \qquad \sum_{i=1}^{N} w_i r_i^{s-1} \, u(-a \log r_i) = \frac{1}{a} F(1/a) \qquad .$$

The values of $u(t)$ occur at $t_k = -a \log r_k$ but the coefficient matrix and the inverse are the same as before.

We can also take advantage of the shifting property of the Laplace Transform. Suppose that we have determined $u(t)$ as a sufficient set of points within the interval $(0, t_0)$, then, using interpolating techniques, we may suppose that $u(t)$ is known within the interval. Then we can write,

$$(6.10.4) \qquad F(s) = \int_0^\infty e^{-st} \, u(t)dt = \int_0^{t_0} e^{-st} \, u(t)dt$$

$$+ \int_{t_0}^\infty e^{-st} \, u(t)dt$$

or

$$(6.10.5) \qquad F(s) - \int_0^{t_0} e^{-st} \, u(t)dt = e^{-st_0}\int_0^\infty e^{-st} \, u(t+t_0)dt \qquad .$$

Evaluating the integral

$$\int_0^{t_0} e^{-st} \, u(t)dt$$

by means of a quadrature formula, we proceed to obtain the values for $u(t+t_0)$ on the interval $(0_0, t_0)$ and so on.

This method requires a high degree of accuracy in the determination of $F(s)$. In many cases this is easy to obtain.

Problems

1. Show that $w_i = \int_0^1 \dfrac{p^*_N(r)dr}{(r-r_i)p^*_N(r_i)}$.

    How does one show that w_i is positive?

2. With the foregoing determination of r_i and w_i, show that

    $$\int_0^1 u(t)dt = \sum_{i=1}^N w_i u(t_i)$$

    for all polynomials of degrees $2N-1$ or less.

3. Consider the method of successive approximations

    $$x_{n+1} = (A^TA + \lambda I)^{-1}(A^Tb + \lambda x_n)$$

    $$x_0 = c .$$

    Show that x_n converges to x for $\lambda > 0$ and any c.

4. In Problem 3 what is the rate of convergence? How does this rate of convergence affect the choice of λ?

5. Let $f(t) = \sin at$, show by contour integration that

    $$F(s) = \int_0^\infty e^{-st} \sin at\, dt = \frac{2}{a^2 + s^2} .$$

6. Compute the first 6 shifted Legendre polynomials. Show that all roots are real, distinct and lie within the interval $|0, 1|$; demonstrate that they are mutually orthogonal.

7. Consider the polynomial $p^*_4(r)$. Demonstrate that the matrix $(w_i r_i^k)$ is ill-conditioned by showing that the value of the determinant is small. Can the process be generalized?

8. Noting that the Legendre polynomial $P_N(r)$ satisfies the second-order ordinary differential equation

$$(1 - r^2)p''_N(r) - 2rp'_N(r) + N(N+1)p_N(r) = 0 \quad ,$$

what is the differential equation for $p^*_N(r)$?

9. Discuss the possibility of numerically integrating this differential equation in the interval $[0, 1]$ to obtain the zeros of $p^*_N(r)$.

10. If two vectors are nearly parallel, their point of inter-section will vary greatly with small changes in either vector. Show that the problem of finding the intersection point leads to solving two simultaneous algebraic equations whose coupling matrix is ill-conditioned.

SUBJECT INDEX